CANCER AND
INFLAMMATION

Novartis Foundation Symposium 256

CANCER AND INFLAMMATION

2004

John Wiley & Sons, Ltd

Published in 2004 by John Wiley & Sons Ltd,
                    The Atrium, Southern Gate,
                    Chichester PO19 8SQ, UK

        National   01243 779777
        International (+44) 1243 779777
        e-mail (for orders and customer service enquiries): cs-books@wiley.co.uk
        Visit our Home Page on http://www.wileyeurope.com
                        or http://www.wiley.com

This publication is designed to provide accurate and authoritative information in regard to
the subject matter covered. It is sold on the understanding that the Publisher is not engaged
in rendering professional services. If professional advice or other expert assistance is
required, the services of a competent professional should be sought.

Other Wiley Editorial Offices

John Wiley & Sons Inc., 111 River Street, Hoboken, NJ 07030, USA

Jossey-Bass, 989 Market Street, San Francisco, CA 94103-1741, USA

Wiley-VCH Verlag GmbH, Boschstr. 12, D-69469 Weinheim, Germany

John Wiley & Sons Australia Ltd, 33 Park Road, Milton, Queensland 4064, Australia

John Wiley & Sons (Asia) Pte Ltd, 2 Clementi Loop #02-01, Jin Xing Distripark, Singapore
129809

John Wiley & Sons Canada Ltd, 22 Worcester Road, Etobicoke, Ontario, Canada M9W 1L1

Wiley also publishes its books in a variety of electronic formats. Some content that appears
in print may not be available in electronic books.

Novartis Foundation Symposium 256
x+280 pages, 32 figures, 13 tables

British Library Cataloguing in Publication Data

A catalogue record for this book is available from the British Library

ISBN 0 470 85510 X

Typeset in 10½ on 12½ pt Garamond by Dobbie Typesetting Limited, Tavistock, Devon.
Printed and bound in Great Britain by T. J. International Ltd, Padstow, Cornwall.
This book is printed on acid-free paper responsibly manufactured from sustainable forestry,
in which at least two trees are planted for each one used for paper production.

Contents

Participants

Fran Balkwill   Cancer Research UK, Translational Oncology Laboratory, Bart's and The London Queen Mary's Medical School, Charterhouse Square, London EC1M 6BQ, UK

Thomas Blankenstein   Max-Delbrück-Centrum, Robert-Rössle-Str 10, 13092 Berlin, Germany

Fionula Brennan   Kennedy Institute of Rheumatology, Faculty of Medicine, Imperial College, 1 Aspenlea Road, Hammersmith, London W6 8LH, UK

Christophe Caux   Schering-Plough Laboratory for Immunological Research, 27 Chemin des Peupliers, B P 11, 69571 Dardilly, France

Vincenzo Cerundolo   Molecular Immunology Group, Institute of Molecular Medicine, John Radcliffe Hospital, Headington, Oxford OX3 9DU, UK

Maurizio D'Incalci   Department of Oncology, Istituto di Ricerche Farmacologiche Mario Negri, via Eritrea 62, 20157 Milan, Italy

Marc Feldmann   Kennedy Institute of Rheumatology, Faculty of Medicine, Imperial College, 1 Aspenlea Road, Hammersmith, London W6 8LH, UK

Guido Forni   Department of Clinical & Biological Sciences, University of Turin, Ospedale San Luigi Gonzaga, Reg. Gonzole 10, 10043 Orbassano, Italy

Awen Gallimore   Department of Medical Biochemistry, University of Wales College of Medicine, Heath Park, Cardiff CF14 4XN, UK

Siamon Gordon   *(Chair)* Sir William Dunn School of Pathology, University of Oxford, South Parks Road, Oxford OX1 3RE, UK

Thorsten Hagemann   *(Novartis Foundation Bursar)* Department of Haematology & Oncology, University of Göttingen, Robert-Koch-Str 40, D-37075, Göttingen, Germany

Adrian Harris   Cancer Research UK, Medical Oncology Unit, Churchill Hospital, Oxford OX3 7LJ, UK

Ian Hermans   The Nuffield Department of Clinical Medicine, John Radcliffe Hospital, Headington, Oxford, OX3 9DS, UK

Alberto Mantovani   Mario Negri Research Institute, Department of Immunology & Cell Biology, Istituto di Richerche Farmacologiche Mario Negri, via Eritrea 62, 20157 Milan, Italy

Robert Cozens   Novartis Pharma AG, CH-4002 Basel, Switzerland

Joost Oppenheim   National Cancer Institute-Frederick Cancer Research and Development Center, Building 560, Room 21-89A, Frederick, MD 21702-1201, USA

Michael S. Pepper   Department of Morphology, University Medical Centre, 1 Rue Michel Servet, 1211 Geneve, Switzerland

Jeffrey W. Pollard   Director, Center for Study of Reproductive Biology and Women's Health, Department of Developmental & Molecular Biology, Albert Einstein College of Medicine, New York, NY 10461, USA

Ann Richmond   Vanderbilt University School of Medicine, 1161 21st Ave South, Nashville, TN 37232, USA

Ari Ristimäki   Department of Pathology, Helsinki University Central Hospital, PO Box 140, FIN-00029, Helsinki, Finland

Barrett Rollins   Department of Medical Oncology, Dana-Farber Cancer Institute, M430, 44 Binney Street, Boston, MA 02115, USA

John Smyth   Director, University of Edinburgh, Cancer Research Centre, Crewe Road South, Edinburgh EH4 2XR, UK

Robert M. Strieter   Division of Pulmonary & Critical Care Medicine and 900 Veteran Avenue, 14-154 Warren Hall, Box 711922, Los Angeles, CA 90024-1922, USA

Michael J. Thun   Epidemiology and Surveillance Research, American Cancer Society, 1599 Clifton Road, NE, Atlanta, GA 30329-4251, USA

Philippe van Trappen   Cancer Research UK, Translational Oncology Laboratory, Bart's and The London Queen Mary's Medical School of Medicine & Dentistry, 3rd Floor, John Vane Science Centre, Charterhouse Square, London EC1M 6BQ, UK

Adam Wilkins   BioEssays, 10/11 Tredgold Lane, Napier Street, Cambridge, CB1 1HN, UK

Chair's introduction

Siamon Gordon

Sir William Dunn School of Pathology, University of Oxford, South Parks Road, Oxford OX1 3RE, UK

The subject of this meeting is actually slightly broader than the narrow focus indicated by its title: we are not just referring to cancer in a specific way, but we are going to include all kinds of tumours, including benign ones. This is also a funny kind of inflammation that we are talking about. It is a modified, mostly chronic rather than acute form, and includes variants of the traditional infectious disease type of inflammation.

Someone yesterday said to me that they thought this meeting could be a turning point in the history of the field. The ideas themselves go back to Virchow and even before, but this subject has never really taken root because of the dominant emphasis on the tumour. The point of this meeting is to put the emphasis on the host response to the tumour.

I have given very little attention to the tumour except to make one or two obvious generalizations. First of all, we think of tumours as either gaining or losing genetic information. With the recent emphasis on tumour suppressors, it has predominantly become one of loss of information. However, there are still examples of viral gene products initiating tumorigenesis. This process usually involves multiple events leading eventually to the full malignant state. Some of the changes that are relevant to the host response include changes in the MHC class I, especially in the non-classical MHC class I molecules. Some of these are not normally expressed, but only as a response to cellular stress. Stress could be very general, but it could also involve all kinds of other pressures from immune cells. There is a lot of interest in this area.

The tumour may express various antigens that are recognizable by the host. These could be complex changes in carbohydrates, lipids and proteins, such as the MUC1 antigen. Of course, the tumour cell is able to express not only surface molecules but also products such as proteolytic enzymes and chemokines.

What about the host interaction with the tumour? Again, I have just listed some of the processes very briefly. We can think of recruitment of host cells, and the formation of a stroma by the host. This is a complex structure that is rather neglected, which represents an attempt to encapsulate a new tumour. An important aspect we will hear a lot about is angiogenesis, as well as lymphatic changes. Then we will hear a bit about a change in the tumour microenvironment, which involves oxygen tension and pH, and then the ability of the tumour to spread through lymphatics and the bloodstream. All of these processes are complex and will involve adjacent fibroblasts, extracellular matrix, macrophages, epithelial cells and the endothelium.

The interactions of the tumour with the host also involve apoptosis of tumour cells, both naturally and chemically induced. This will determine to some extent its ability to invade. The host may assist in this or inhibit it. Proteolysis is an important theme here. One theme that everyone now accepts is that the host response may provide a trophic role for the tumour and promote the emergence of mutant clones, but it also has the potential of being cytotoxic and being able to eliminate tumours. The tumour naturally can invade some of these responses and even block them actively. These effects may be expressed through antigen-presenting cells. We will hear about dendritic cells as well as macrophages, and also regulatory T cells. As a result of this, the tumour — which is genetically unstable — can select for variants that are resistant to host responses. This creates considerable problems for the host. The infectious disease model for tumour–host interactions is actually quite a useful principle. We have learnt a lot from studying virus infections as well as bacterial infections.

One or two words about the macrophages and their relationship to tumours. They are important because of their long-lived nature, they are biosynthetically extremely active, and they are present in all the tissues of the body. They themselves can provide both trophic and cytotoxic functions. They are considerably heterogeneous. There are resident cells in the tissues in the absence of inflammation, and many are recruited in response to inflammatory or infectious stimuli, but also they are immunologically activated by certain pathogens, and perhaps deactivated by the tumour. The cells can make contact and therefore interact through plasma membrane interactions, as well as through local secretion and humoral, systemic release of products. Of course, their prime function of professional phagocytic activity is a characteristic feature of many tumours.

There is one category of molecule we are not going to discuss much at this meeting (partly because it has hardly been studied) characterized by the ability of macrophages and other antigen-presenting cells (APCs) to use their so-called 'pattern recognition receptors', which are on the whole non-opsonic. Of course antibody and complement, as opsonic receptors, contribute to the uptake under certain circumstances. But in many cases there is no particular antibody that we can identify, or fixation of complement. Thus these other non-opsonic receptors become of major interest. The host depends on them and they are germline-encoded. They don't rearrange like T and B cell receptors. They are not clonally specific, but instead see a conserved structural pattern on a particular target. This concept has arisen in terms of microorganisms, but I think it can be extended to tumour recognition also. These receptors are not only directed against exogenous ligands, but also against endogenous ligands, which are either the apoptotic cells themselves, recognized by similar receptors, or by modified self components. This is an important theme of current research. It doesn't have to be truly foreign in the way bacteria might be; it could be something that is denatured or altered so that the host knows that this homeostatic cell — the macrophage — has to remove this to maintain homeostasis. An important family of signalling components, the toll-like receptors, are under intense scrutiny. These can be stimulated by endotoxin, which results in signalling by NF-κB activation. Many of the pro-inflammatory cytokines that we talk about, such as tumour necrosis factor (TNF), are primarily regulated by this signalling pathway in response to selected recognition processes. There are also other receptors which down-regulate macrophage responses. One family that we are interested in is the epidermal growth factor (EGF) TM7 family. There are other newly described molecules such as CD200 and its receptor, and CD47. People are beginning to realize that these are balanced processes: for every system that is induced, there must be down-regulatory or inhibitory systems as well.

A final word about the macrophage and its activation spectrum. The classical form of activation is through gamma interferon produced by NK cells and CD4/CD8 lymphocytes. This is what is responsible for classic cell-mediated immunity to intracellular pathogens and the ability to kill target cells as well as organisms. The opposite of this is deactivation by interleukin (IL)10, transforming growth factor (TGF)β and colony-stimulating factor (CSF)-1 (also known as M-CSF). This is able to counteract inflammatory responses which are mediated by certain cytokines. It is anti-inflammatory and can suppress or prevent some of the responses that are associated with infection. In between there is a form of activation that we have termed alternative activation, induced by IL4 and IL13 acting through a common receptor-α chain. Alternative activation of macrophages is responsible for induction of some immune responses, especially humoral immunity, to extracellular pathogens and parasites, and possibly to tumours, as well as to other forms of host alteration. These responses seem to play a key role in the repair process, so that the interactions between macrophages and fibroblasts, for example, which are still poorly defined, may well involve this kind of cell activation.

I would like to briefly describe EGF TM7 molecules. Some 20 years ago we isolated a monoclonal antibody called F4/80, directed against an antigen which has become a prototypical marker of mature resident macrophages in tissues of mice. When the protein recognized by this antibody was cloned, we found that it had a seven transmembrane portion that was highly homologous to a family of G protein-coupled peptide receptors. Unlike these peptide receptors, it has a very large extracellular domain consisting of seven EGF modules and a stalk. The family has grown in recent years. The best known example is probably CD97. All of these are found on monocytes and macrophages, and CD97 also more broadly on some non-haematopoietic cells as well, including fibroblasts and smooth muscle. There is a very similar organization among all family members, with different numbers and sequences in the EGF modules. CD97 was the only one previously shown to have a known ligand, CD55, a complement regulatory protein. It appeared that in order to signal through the 7TM portion, and possibly set up inhibitory signals rather than activating signals, there might be specific recognition of ligands in matrix and on fibroblasts. We looked for ligands in tissues, and we have found specific ligands for EMR2. The strategy we used is a powerful one. The interactions between these receptors and their ligands tend to be low affinity, so in order to make a detectable system we made a multivalent high-avidity probe. We made a chimeric protein with human or mouse Fc, then multimerized these on streptavidin B-coated beads that could be fluorescent or magnetic. This provides a tool to examine cell lines by, for example, fluorescence-activated cell sorting (FACS) analysis. By using this strategy we have found that dermatan sulfate provides ligands for this particular member of the EGF TM7 family. This proteoglycan is found in tumour matrix. We have three way interactions potentially, between the tumour cell, macrophage and fibroblast. The receptor is G protein coupled and can send a signal into the macrophage, and through the dermatan sulfate it can send a signal into the adherent cell, if it is a fibroblast, or fail to do so if it is a tumour cell that is not responding to some of these signals. We believe that interactions between these macrophage surface molecules and tumour and fibroblast molecules provide a new way for reciprocal regulation of cell migration, adhesion and potentially to be retained at tissue sites. Cellular interactions can contribute not only to recognition of these cells, but may provide a kind of nursing function and possibly contribute to cytotoxic activities. It is all pretty open-ended still: we don't have much evidence that this actually happens, but I believe it is relevant to what happens in the tumour stroma. As a result of this meeting I hope to find some good models to study this particular hypothesis.

What are the questions that we should be thinking of? We need to decide what are suitable experimental models for both *in vivo* and *in vitro* analysis. The tumour and the host have many changes both in their genetic programmes and epigenetic regulation of some of the genes that are expressed in both components. From the point of view of immune recognition and manipulation, the tumour might mask its antigens on the surface, or we may find ways of enhancing recognition. Of course, in terms of generating an immune response, many of the adjuvants that are under study for vaccine production might be highly pertinent for making a tumour vaccine as well. Some monoclonal antibodies have turned out to be very powerful anti-tumour agents. Finally, what are the factors that control the balance between the trophic and cytotoxic interactions of the host? I'm sure there are many other important questions that will be addressed over the next few days.

Inflammation and cancer: an epidemiological perspective

Michael J. Thun, S. Jane Henley and Ted Gansler

American Cancer Society, 1599 Clifton Road, Atlanta, GA, 30329-4251, USA

Abstract. Many chronic inflammatory conditions increase the risk of cancer in affected tissues. Clinical conditions that involve both inflammation and increased cancer risk include a broad range of immunological disorders, infections (bacterial, helminthic, viral), and chronic chemical and mechanical irritation. For example, the inflammatory bowel diseases, ulcerative colitis and Crohn's disease, predispose to the development of cancers of the large bowel and/or terminal ileum; chronic infection with the bacterium *Helicobacter pylori* causes atrophic gastritis, dysplasia, adenocarcinoma and an unusual form of gastric lymphoma; and parasitic infection with schistosomes and other trematodes cause cancers of the urinary bladder and the intrahepatic and extrahepatic biliary tract. Chronic reflux of gastric acid and bile into the distal oesophagus causes chemical injury, Barrett's oesophagus and oesophageal adenocarcinoma. Chronic cholecystitis and gallstones predispose to cancer of the gallbladder. Besides these clinical syndromes, subclinical inflammation may promote the development of certain tumours. The expression of COX-2 and lipid mediators of inflammation increases during the multistage progression of these tumours. Non-steroidal anti-inflammatory drugs (NSAIDs), which inhibit COX-2 activity and tumour development in many experimental and clinical settings, are inversely associated with certain cancers in epidemiological studies. Despite their promise, however, anti-inflammatory drugs are not yet recommended for the prevention or treatment of any cancers. Numerous questions must be resolved concerning their molecular and cellular targets of action, efficacy, safety, treatment regimen, indications, and the balance of risks and benefits from treatment in designated patient populations.

2004 Cancer and inflammation. Wiley, Chichester (Novartis Foundation Symposium 256) p 6–28

The hypothesis that chronic irritation or injury may predispose to the development of certain cancers was raised by Virchow in the mid-19th century (Parsonnet 1999). He theorized that chronic irritation may establish the setting in which cells grow abnormally, as exemplified by bladder cancer occurring in patients from North Africa infected with *Schistosoma haematobium* (Parsonnet 1999). Numerous case reports and clinical series have described carcinomas of the skin arising as a complication of burns and scars, chronic sinus tracts, fistulas (Kaplan 1987,

Scotto et al 1996) and ulcers (Parsonnet 1999). Lung carcinomas have been reported at the site of scar tissue in patients with previous tuberculosis (Auerbach et al 1979) and sarcomas can occur as a complication of surgical implants and foreign bodies (IARC 1999). However, the evidence implicating many of these chronic inflammatory conditions with cancer is limited. Much of the information derives from case reports. Scar tissue that adjoins a carcinoma may be a consequence rather than a cause of tumour growth (Blot & Fraumeni 1996). In any event, the great majority of cancers that occur arise in patients and tissues with no obvious chronic inflammatory disease.

A more contemporary version of Virchow's hypothesis is that the inflammatory processes induced by chronic injury contribute to the multistage development of cancer and that these, rather than the specific cause of the injury, account for the carcinogenicity in the majority of settings listed above. Inflammation involves a complex of host responses that, in the context of acute injury, promote wound healing and tissue regeneration. These responses include recruitment of specific types of cells, release of inflammatory mediators and interactions among chemokine ligand/receptor systems. Leukocytes (neutrophils, monocytes, macrophages, and eosinophils) generate reactive oxygen and nitrogen species that can directly damage the genes that control cell growth (Christen et al 1999). Cells that mediate the inflammatory response release autocrine and paracrine factors that stimulate cell proliferation, inhibit apoptosis, induce angiogenesis, and impair certain immune responses. Collectively, these factors can accelerate mutagenesis, promote the survival and clonal proliferation of mutated cells, and increase the probability that a particular clone of cells will acquire the requisite genetic mutations to become an invasive and metastatic cancer.

Even in the absence of overt inflammation, many of the same factors that mediate the acute inflammatory response are also produced by solid tumours at various stages of their development. For example, factors that stimulate cell proliferation, inhibit apoptosis and induce angiogenesis are involved in both wound healing and carcinogenesis. Enzymes like the inducible form of cyclooxygenase (COX-2) are expressed during wound healing and certain stages of neoplasia and increase production of inflammatory mediators. Much of the ongoing research on inflammation and cancer now focuses on the potential role of subclinical inflammatory mediators on the development of a wide range of cancers rather than on clinical inflammatory conditions known to predispose to specific cancers.

This overview considers three lines of evidence that are relevant to the hypothesis that chronic inflammation promotes the development of certain cancers. First it describes the broad spectrum of clinical disorders that involve both chronic inflammation and increased cancer risk. These include chronic inflammation from certain immunological conditions (ulcerative colitis, Crohn's disease, etc.), from chemical or mechanical irritation (reflux oesophagitis, gall stones), and from selected infections (bacterial, helminthic and viral) (Shacter & Weitzman 2002). Next it considers the increased expression of various inflammatory mediators that occurs during the development of various tumours. Finally, it considers the epidemiological, clinical and experimental evidence that non-steroidal anti-inflammatory drugs (NSAIDs) inhibit the occurrence or progression of certain cancers (Thun et al 2002).

Clinical conditions that involve chronic inflammation and cancer

Clinicians have long been aware that a wide variety of chronic inflammatory disorders predispose to malignancy in the affected organ(s). Table 1, modified from Shacter & Weitzman (2002), lists examples of chronic inflammatory diseases that give rise to carcinomas and/or sarcomas. The only haematopoeitic cancer mentioned in Table 1 is the mucosa-associated lymphoid tumour (MALT), an unusual lymphoma that, like gastric adenocarcinoma, can be induced by chronic infection with *Helicobacter pylori*. Other conditions that predispose to lymphoma through chronic immune stimulation, and viral agents thought to induce cancer through direct interactions with host DNA, are not included in Table 1.

Idiopathic immunologically mediated conditions

Inflammatory bowel disease (IBD).   Ulcerative colitis and Crohn's disease are related but clinically and histologically distinct inflammatory diseases of the bowel that predispose to adenocarcinomas of the large bowel, terminal ileum, and in some cases extraintestinal sites including the biliary tract (Podolsky 2002). Tumours often occur at the site of chronic inflammation (Fenkel 2002). In ulcerative colitis the intestinal inflammation is limited to the colon and rectum, whereas in Crohn's disease, two-thirds of patients have inflammation and increased risk of cancer in the terminal ileum. Inflammation is thought to result from a combination of genetic susceptibility and inappropriate activation of the mucosal immune system by normal luminal flora (Podolsky 2002). The absolute risk of developing colorectal cancer is high when extensive disease begins at a young age. Forty percent of patients diagnosed with pancolitis from ulcerative colitis before age 15 years developed colon cancer during 20-years of follow-up in a population-based linkage study in Sweden (Ekbom et al 1990). Numerous specific and non-specific inflammatory factors are expressed in patients with these conditions. These include phagocytic products (oxygen metabolites, nitric oxide, collagenases, etc.), toxic lymphocyte products, cytokines (including chemokines), neuropeptides and various components of plasma proteolytic cascades (Podolsky 2002).

TABLE 1   Chronic inflammatory conditions that predispose to cancer

Immunologic conditions	Cancer	Aetiologic agent
Ulcerative colitis	Colon	Immune response
Crohn's disease	Colon and terminal ileum	Immune response
Primary sclerosing cholangitis	Cholangiocarcinoma	Immune response
Lichen sclerosis	Vulvar carcinoma	
Oral lichen planus	Oral cavity	
Autoimmune gastritis	Gastric adenocarcinoma	Immune response
Chemical and mechanical irritation		
Reflux oesophagitis Barrett's oesophagus	Oesophagus, adenocarcinoma	Gastric acid and bile
Familial pancreatitis	Pancreas	Cationic trypsinogen
Sporadic chronic pancreatitis	Pancreas	Pancreatic enzymes
Haemochromatosis	Liver	Haemosiderin
Gallstones	Gall bladder	Chronic cholecystitis
Silicosis	Lung cancer	Silica
Asbestosis	Mesothelioma, lung cancer	Asbestos
Infectious conditions BACTERIA		
Chronic gastritis	Gastric adenocarcinoma, MALT	*Helicobacter pylori*
Chronic osteomyelitis	Skin	Various species
Catheter-associated cystitis	Urinary bladder	Various species
Cervicitis	Cervix	*Chlamydia trachomatis*
HELMINTHS		
Cystitis	Urinary bladder	*Schistosoma haematobium*
Cholangitis	Biliary tract	*C sinensis, O viverinni*
VIRUSES		
Chronic viral hepatitis	Liver	Hepatitis B and C

Other immunological diseases.   Various other idiopathic immunologically-mediated conditions that predispose to certain cancers are listed in Table 1. These include primary sclerosing cholangitis, lichen sclerosis, oral lichen planus and autoimmune gastritis. The list is not intended to be inclusive. Not mentioned are Hashimotos's thyroiditis or Sjogren's syndrome, which are associated with lymphomas but not solid cancers (Ron 1996). Lymphomas may arise from chronic polyclonal stimulation of immune cells leading to monoclonal proliferation rather than from more general inflammatory processes.

Chemical and mechanical irritation

Gastroesophageal reflux. Chronic reflux of gastric acid and bile from the stomach into the distal oesophagus causes chemical inflammation and histological abnormalities known as Barrett oesophagus (Shaheen & Ransohoff 2002). Damaged squamous cell epithelium of the lower oesophagus is replaced by metaplastic intestinal-type columnar epithelium that in some patients progresses to high-grade dysplasia. The risk of developing adenocarcinoma is estimated to be 0.5% per year among all patients with Barrett oesophagus, but 25% among those with high-grade dysplasia (Shaheen & Ransohoff 2002). Aggressive treatment with proton pump inhibitors has not been shown to induce regression of the histological abnormalities. The incidence of oesophageal adenocarcinoma has increased more rapidly than that of any other cancer in the USA since the mid-1970s (Devesa et al 1998).

Familial pancreatitis, sporadic chronic pancreatitis and haemachromatosis all cause chemically induced chronic inflammation with increased risk of cancer in the affected organ.

Gallstones. Chronic cholecystitis from recurrent or persistent gallstones predisposes to biliary tract cancer. The risk of gallbladder cancer is reportedly 4–5 times higher in patients with than without gallstones (Lowenfels et al 1999). Factors that predispose to gallstone formation are also risk factors for biliary tract cancer. These include obesity, multiple pregnancies and a genetic disorder of cholesterol metabolism prevalent among indigenous populations of North and South America.

Pneumoconioses. Asbestos fibres deposited in the lung and pleura cause chronic inflammation, asbestosis (pleural plaques and interstitial fibrosis), and increased risk of both mesothelioma and lung cancer (IARC 1977). Exposure to crystalline silica causes both pulmonary fibrosis and increased risk of lung cancer (IARC 1997).

Surgical implants. Metal nails and implants, as in hip replacement, are occasionally associated with cancers of the bone and other adjacent tissues at the site of the implant (IARC 1999). This literature consists largely of case-reports rather than systematic epidemiological studies.

Infectious conditions

Bacterial — Helicobacter pylori. Clinical, epidemiological, and experimental studies have established that the bacterium *H. pylori* is the principal cause of both gastric adenocarcinoma and of an uncommon mucosa-associated lymphoid tissue

(MALT) lymphoma of the stomach (IARC 1994a, Suerbaum & Michetti 2002). *H. pylori* is a Gram-negative, spiral bacteria that colonizes the mucus layer of the stomach. Its pathogenicity is greatest when colonization begins in childhood and involves more virulent strains that express the CagA (a 128 kDa cytotoxin-associated gene A-positive) protein. Adverse effects are also modified by host factors such as diet and genetic susceptibility (Yamaguchi & Kakizoe 2001). Persistent infection causes the development of chronic atrophic gastritis, achlorhydria, intestinal metaplasia, dysplasia and adenocarcinoma in susceptible persons. Similar findings are seen in experimental studies of ferrets and rhesus monkeys infected with other species of *Helicobacter* (Nightingale & Gruber 1994). Premalignant gastric lesions (Correa et al 2000) and low grade B cell gastric lymphomas (Wotherspoon et al 1993) may regress following successful eradication of the infection. Inflamed gastric epithelium is a rich source of interleukin 8 (IL8) and epithelial-cell-derived neutrophil-activating peptide 78 (Suerbaum & Michetti 2002). Cytotoxic strains of *H. pylori* that produce CagA+ induce greater production of IL8 from gastric epithelial cells.

Bacterial other. Other examples of chronic bacterial infection that may increase cancer risk include chronic osteomyelitis (predisposing to cancers of the skin and bone), chronic draining fistulas (causing local squamous cell carcinomas of the skin), chronic indwelling catheters (causing cystitis and bladder cancer) and cervicitis from *Chlamydia*.

Helminth infections — Schistosoma haematobium. Chronic infection with *Schistosoma haematobium* accounts for a substantial fraction of cancers of the urinary bladder in Egypt, where bladder cancer comprises approximately one-third of all cancers in men (Ferlay et al 2000). Many of the severe pathological manifestations of schistosomiasis such as ulcers, bladder polyps, fistulae, and strictures result from the physical and immunological response of the host to the eggs rather than direct effects of the organism (Rosin & Hofsath 1999). The eggs from *S. haematobium* are deposited largely in the terminal ureters and bladder (IARC 1994a). Egg deposition stimulates the activation and recruitment of monocytes and other leukocytes, which contribute to chronic ulceration. Egg remnants have been demonstrated in 82% of patients with urinary bladder cancer from *S. haematobium* in one large case series in Egypt (El-Biokany et al 1981). The bladder tumours associated with *S. haematobium* are typically squamous cell rather than transitional carcinomas. There is some evidence implicating another species of schistosome, *S. japonicum* in colorectal cancer (Rosin & Hofseth 1999), although the International Agency for Research on Cancer (IARC) has classified this evidence as 'limited' (IARC 1994a).

Infection with other trematodes.   Three species of liver fluke infest the intrahepatic bile ducts of people who consume raw fish in parts of Southeast Asia (*Opisthorchis viverrini*), China and neighbouring countries (*Clonorchis sinensis*), and the former Soviet Union and Eastern Europe (*O. felineus*). These trematodes attach themselves to the intra- and extra-hepatic bile ducts where the adult parasite survives for 25–30 years (Thamavit et al 1999). All three species cause chronic inflammation and fibrosis, but only *O. viverrini* has been studied systematically and classified as a definite human carcinogen by IARC (IARC 1994a). Chronic inflammation is thought to cause progression from metaplasia, to dysplasia and cholangiocarcinoma.

Viral infections.   Most hepatocellular carcinomas (HCC) in persons with chronic hepatitis B or C infection occur in conjunction with cirrhosis or chronic hepatitis (IARC 1994b). Recurrent cycles of inflammation, necrosis and regeneration appear more important to the carcinogenicity of HBV and HCV than are direct effects of the virus on host DNA (IARC 1994b). Infection acquired in early childhood confers the highest risk of cirrhosis and hepatocellular carcinoma. Prospective studies have shown that the incidence of HCC is more than 100-fold greater in patients with chronic HBV infection than in uninfected individuals (Robinson 1999).

Other viruses known to cause cancer include human papilloma virus (cervix, vulva, anus, penis, possibly oropharynx and oesophagus), Epstein Barr virus (lymphoma, nasopharynx), Kaposi's sarcoma-associated herpes virus, and human T-cell leukaemia viruses (HTLVs). Although inflammation occurs at certain stages in the development of these tumours, it is difficult to separate the direct oncogenic effects of these viruses on DNA from their secondary effects on the immune system or from pathways involving chronic inflammation.

Insights regarding clinical inflammation and cancer

In summary, a broad range of human disorders cause chronic inflammation and predispose to increased risk of cancer in the affected organ(s). These provide strong observational support for the hypothesis that chronic inflammation contributes causally to the development of at least some of these tumours, but do not prove the hypothesis. Researchers have not yet identified specific therapeutic targets that inhibit tumour development in these settings. In some conditions, such as chronic viral infections, the inflammatory processes cannot be completely separated from the direct effects on DNA caused by the underlying pathogen. In others, such as inflammatory bowel disease, reflux oesophagitis and mechanical irritation from gallstones, the inflammatory response itself is clearly the principal determinant of cancer risk. These disorders, particularly the conditions that are not amenable to antibiotic therapy, provide clinical opportunities to assess mechanism and to conduct randomized clinical trials.

Diseases that involve persistent clinical inflammation differ in severity from conditions in which inflammatory mediators may be expressed at subclinical levels in one or more stages of the development of certain cancers. However, the mechanisms by which chronic inflammation affects the development of human cancer may be the same. All of these conditions involve factors that may stimulate cell proliferation, suppress apoptosis, induce angiogenesis and disrupt other aspects of host immunity. These clinical settings provide opportunities to examine the role of specific cytokines in initiating or maintaining the inflammatory response and the role of reactive oxygen and nitrogen species in causing mutations. They also provide opportunities for randomized clinical trials to determine whether anti-inflammatory drugs prevent the occurrence or progression of neoplasia.

Expression of inflammatory mediators during tumour development

A second line of evidence concerns the increased expression of inflammatory mediators that occurs during tumour development. Numerous studies have shown that the inducible isoform of cyclooxygenase, COX-2, is over-expressed during the development of many cancers (reviewed in Gupta & DuBois 2001, Masferrer et al 2000). COX-2 is the inducible isoform of cyclooxygenase, the rate-limiting enzyme that converts arachidonic acid to prostaglandins and other metabolites. Increased expression of COX-2 has been documented in biopsies from colorectal adenomas and from carcinomas of the colon, stomach, oesophagus, pancreas, bladder, skin (non-melanoma), lung, head and neck and from melanoma (reviewed in Thun et al 2002, Masferrer et al 2000). The increased expression of COX-2 in many human tumours has stimulated numerous clinical and experimental studies that demonstrate that tumour growth can be inhibited by COX-2 inhibiting drugs and/or by genetic knockout of COX-2 activity (Thun et al 2002). Collectively, these studies provide strong evidence that lipid mediators of inflammation contribute to the multistage development of multiple cancers.

Although COX-2 is an attractive target in studies of chronic inflammation and cancer because of the availability of NSAIDs that selectively block COX-2 activity, many questions remain about the role of COX-2 in tumour development. Researchers have not yet identified the mechanism by which lipid mediators modulate apoptosis and angiogenesis in various experimental models, nor the relevance of these models for human cancers. Furthermore, the intensity of COX-2 expression varies widely across different types of tumours. It is not clear whether and in which clinical situations COX-2 actually contributes to tumour

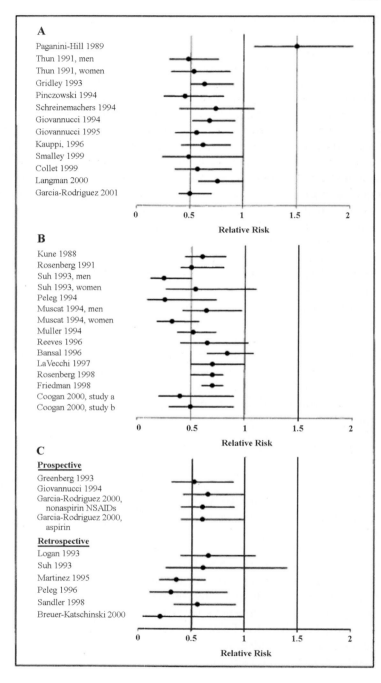

A

Paganini-Hill 1989
Thun 1991, men
Thun 1991, women
Gridley 1993
Pinczowski 1994
Schreinemachers 1994
Giovannucci 1994
Giovannucci 1995
Kauppi, 1996
Smalley 1999
Collet 1999
Langman 2000
Garcia-Rodriguez 2001

Relative Risk

B

Kune 1988
Rosenberg 1991
Suh 1993, men
Suh 1993, women
Peleg 1994
Muscat 1994, men
Muscat 1994, women
Muller 1994
Reeves 1996
Bansal 1996
LaVecchi 1997
Rosenberg 1998
Friedman 1998
Coogan 2000, study a
Coogan 2000, study b

Relative Risk

C

Prospective
Greenberg 1993
Giovannucci 1994
Garcia-Rodriguez 2000,
  nonaspirin NSAIDs
Garcia-Rodriguez 2000,
  aspirin

Retrospective
Logan 1993
Suh 1993
Martinez 1995
Peleg 1996
Sandler 1998
Breuer-Katschinski 2000

Relative Risk

progression or how the effects of COX-2 may vary depending upon the cellular target.

There has been less research on the role of cytokines and other inflammatory mediators in tumour development. Mantovani et al (2002) demonstrated that serum levels of proinflammatory cytokines IL2, IL6, tumour necrosis factor (TNF)α, leptin and C-reactive protein were higher in 82 advanced cancer patients than in 36 controls. The concentration of C-reactive protein in plasma was reported to correlate inversely with survival in cancer patients with multiple myeloma, melanoma, lymphoma and tumours of the ovary, pancreas and gastrointestinal tract (Mahmoud & Rivera 2002). Although cancer researchers have begun to study cytokines as predictors of survival in patients with cancer, there have not yet been studies that measure cytokine concentrations in serum or plasma before the diagnosis of cancer as predictors of incidence. This is surprising, since cardiovascular researchers have now established that circulating levels of C-reactive protein, IL6 and other cytokines are strong and independent predictors of coronary heart disease and stroke (Pradham et al 2002), and research hypotheses regarding cardiovascular and cancer research are often pursued in parallel.

Tumour inhibition by NSAIDs

A third line of evidence that links chronic inflammation in the development of certain cancers involves studies of tumour inhibition by NSAIDs. Randomized clinical trials have established that two NSAIDs, the prodrug sulindac and the selective COX-2 inhibitor celecoxib effectively inhibit the growth of adenomatous polyps and cause regression of existing polyps in patients with the hereditary condition familial adenomatous polyposis (FAP). NSAIDs have been shown to inhibit tumour development, restore apoptosis, and to suppress angiogenesis in a variety of *in vivo* and *in vitro* experimental models (Thun et al 2002).

Numerous epidemiological (non-randomized) studies have found lower incidence of adenomatous polyps and lower incidence or death from colorectal cancer in persons who regularly use aspirin and other NSAIDs compared with non-users (Fig. 1), although one study has not (reviewed in Thun et al 2002). Prolonged use of aspirin or other NSAIDs is consistently associated with a 30–50% reduction in incidence or death rates from colorectal cancer in all but one

---

FIG. 1.   Epidemiological studies of the association between non-steroidal anti-inflammatory drug (NSAID) use and colorectal cancer or adenomatous polyps. The relative risk estimates (circles) and 95% confidence intervals (lines) refer to the incidence or death rates among regular NSAID users compared to that of non-users in (A) cohort studies and (B) case-control studies of NSAIDs and colorectal cancer, and (C) studies of NSAIDs and adenomatous polyps. Reproduced with permission from Thun et al (2002) which includes details of references cited.

of these epidemiological studies (Fig. 1) (Thun et al 2002). Regular NSAID use is also associated with lower risk of cancer of the stomach (Fig. 2) and oesophagus (Fig. 3) in most studies (Thun et al 2002). NSAID use is associated with reduced risk of breast cancer in some studies (Fig. 4), but not in the Nurses' Health Study (Egan et al 1996) nor in the large American Cancer Society Cancer Prevention Study (Thun et al 1993). Howe et al (2002) have proposed that NSAIDs may protect against breast cancer only in the subgroup of tumours that express HER-2/neu. Therefore, future studies should take into account the heterogeneity of tumour types.

Insights from research on NSAIDs and cancer inhibition

Collectively, the clinical, epidemiological and experimental studies provide strong evidence that arachidonic acid or its metabolites affect the development of certain cancers, particularly cancers of the colorectum, stomach, and oesophagus (reviewed in Thun et al 2002). Despite the strengths of these studies, however,

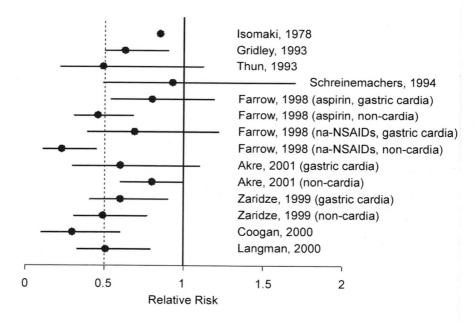

FIG. 2.   Epidemiological studies of the association between non-steroidal anti-inflammatory drug (NSAID) use and stomach cancer. The relative risk estimates (circles) and 95% confidence intervals (lines) refer to the incidence or death rates among regular NSAID users compared to those of non-users.

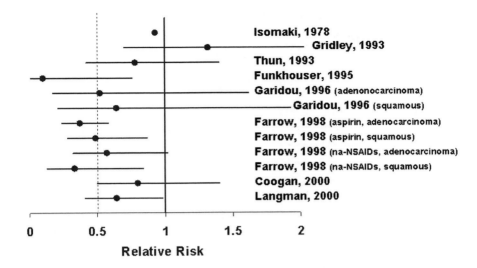

FIG. 3.   Epidemiological studies of the association between non-steroidal anti-inflammatory drug (NSAID) use and oesophageal cancer. The relative risk estimates (circles) and 95% confidence intervals (lines) refer to the incidence or death rates among regular NSAID users compared to those of non-users (na-NSAIDS, non-aspirin NSAIDS).

they do not provide randomized evidence that NSAIDs prevent the development of adenomatous polyps or cancer in the general population. Furthermore, studies have not yet defined the optimal drug, dose, treatment regimen, age to begin prophylactic therapy, or the balance of risks and benefits in different patient populations.

Conclusions

In summary, it is well established that a wide variety of chronic inflammatory diseases give rise to cancer, and that in some instances, the inflammatory response of the host rather than the specific cause of the injury appears to be the principal determinant of cancer risk. Besides these clinical syndromes, subclinical inflammation may promote the development of certain tumours. The concentration of COX-2 and related lipid mediators of inflammation increases during the multistage development of colorectal and other human cancers. It is biologically plausible that chronic inflammation predisposes to cancer, since cells involved in the immune response generate reactive oxygen and nitrogen species that are directly mutagenic, and release autocrine and paracrine factors that stimulate the clonal proliferation of genetically damaged cells. The resultant loss of tumour suppressor function, acquisition of oncogenes, and/or

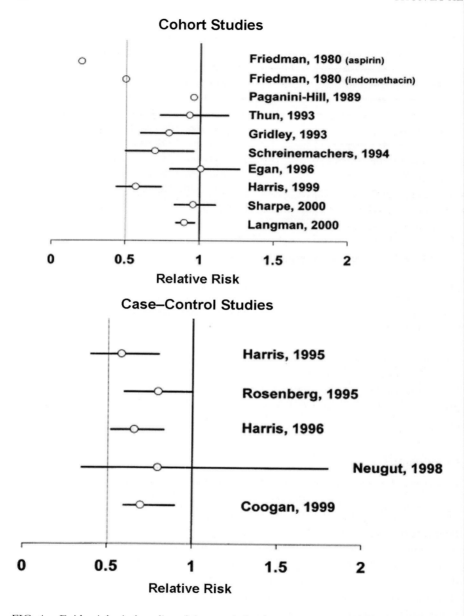

FIG. 4.   Epidemiological studies of the association between non-steroidal anti-inflammatory drug (NSAID) use and breast cancer. The relative risk estimates (circles) and 95% confidence intervals (lines) refer to the incidence or death rates among regular NSAID users compared to those of non-users.

loss of caretaker genes confers a selective growth advantage to mutated cells.

However, numerous questions must be answered before the current interest in inflammation and cancer can be translated into clinically useful therapies for the prevention or treatment of cancer. It has not yet been determined which aspects of the inflammatory process are critical to tumour development and which are incidental. Other questions concern the types of cancer and stages of development for which inflammation is important. Additional specific challenges are:

- to identify the molecular and cellular targets of anti-inflammatory drugs in cancer
- to determine the optimal drug, dose, and treatment regimen
- to assess the safety of treatment and the balance of risks and benefits in specific patient populations and
- to develop guidelines for clinicians and patients that take into account multiple health endpoints in patients with specific clinical profiles.

References

Akre K, Ekstrom A, Signorello L, Hansson L, Nyren O 2001 Aspirin and risk for gastric cancer: a population-based case-control study in Sweden. Br J Cancer 84:965–968

Auerbach O, Garfinkel L, Parks VR 1979 Scar cancer of the lung: increase over a 21 year period. Cancer 43:636–642

Blot WJ, Fraumeni JF 1996 Cancers of the lung and pleura. In: Schottenfeld D, Fraumeni JF (eds) Cancer epidemiology and prevention. Oxford University Press, New York, p 637–665

Christen S, Hagen TM, Shigenaga MK, Ames BN 1999 Chronic inflammation, mutation, and cancer. In: Parsonnet J (ed) Microbes and malignancy—infection as a cause of human cancers. Oxford University Press, New York, p 35–88

Coogan PF, Rao SR, Rosenberg L et al 1999 The relationship of nonsteroidal anti-inflammatory drug use to the risk of breast cancer. Prev Med 29:72–76

Coogan P, Rosenberg L, Louik C et al 2000 NSAIDs and risk of colorectal cancer according to presence or absence of family history of the disease. Cancer Causes Control 11:249–255

Correa P, Fontham ETH, Bravo JC et al 2000 Chemoprevention of gastric dysplasia: randomized trial of antioxidant supplements and anti-Helicobacter pylori therapy. J Natl Cancer Inst 92:1881–1888

Devesa SS, Blot WJ, Fraumeni JF Jr 1998 Changing patterns in the incidence of esophageal and gastric carcinoma in the United States. Cancer 83:2049–2053

Ekbom A, Helmick C, Zack M, Adami H-O 1990 Ulcerative colitis and colorectal cancer. A population-based study. New Engl J Med 323:1228–1233

Egan KM, Stampfer MJ, Giovannucci E, Rosner BA, Colditz GA 1996 Prospective study of regular aspirin use and the risk of breast cancer. J Natl Cancer Inst 88:988–993

El-Bokainy MN, Mokhtar NM, Ghoneim MA, Hussein MH 1981 The impact of schistosomiasis on the pathology of bladder carcinoma. Cancer 48:2643–2648

Farrow DC, Vaughan TL, Hansten PD et al 1998 Use of aspirin and other nonsteroidal anti-inflammatory drugs and risk of esophageal and gastric cancer. Cancer Epidemiol Biomarkers Prev 7:97–102

Ferlay J, Bray F, Pisani P, Parkin DM, GLOBOCAN 2000 Cancer incidence, mortality, and prevalence worldwide, Version 1, IARC CancerBase No. 5, IARC Press, Lyon, France

Frenkel K 2002 The Shacter/Weitzman article reviewed. Oncology 16:230–232

Friedman G, Ury H 1980 Initial screening for carcinogenicity of commonly used drugs. J Natl Cancer Inst 65:723–733

Funkhouser E, Sharp G 1995 Aspirin and reduced risk of esophageal carcinoma. Cancer 76:1116–1119

Garidou A, Tzonou A, Lipworth L, Signorello L, Kalapothak V, Trichopoulos D 1996 Lifestyle factors and medical conditions in relation to esophageal cancer by histologic type in a low-risk population. Int J Cancer 68:295–299

Gridley G, McLaughlin J, Ekbom A 1993 Incidence of cancer among patients with rheumatoid arthritis. J Natl Cancer Inst 85:307–311

Gupta RA, DuBois RN 2001 Colorectal cancer prevention and treatment by inhibition of cyclooxygenase-2. Nat Rev Cancer 1:11–21

Harris R, Namboodiri K, Stellman S, Wynder E 1995 Breast cancer and NSAID use: heterogeneity of effect in a case-control study. Prev Med 24:119–120

Harris R, Namboodiri K, Farrar W 1996 Nonsteroidal antiinflammatory drugs and breast cancer. Epidemiology 7:203–205

Harris R, Kasbari S, Farrar W 1999 Prospective study of nonsteroidal anti-inflammatory drugs and breast cancer. Oncol Rep 6:71–73

Howe LR, Subbaramaiah K, Patel J, Masferrer JL, Deora A, Hudis C 2002 Celecoxib, a selective cyclooxygenase inhibitor, protects against human epidermal growth factor receptor 2 (HER-2)/neu-induced breast cancer. Canc Res 62:5405–5407

IARC 1977 Asbestos. IARC Monographs on the evaluation of carcinogenic risks to humans. Vol 14. IARC Press, Lyon, France, p 106

IARC 1994a Schistosomes, liver flukes, and Helicobacter pylori. IARC Monographs on the evaluation of carcinogenic risks to humans. Vol 61. IARC Press Lyon, France, p 270

IARC 1994b Hepatitis viruses. IARC Monographs on the evaluation of carcinogenic risks to humans. Vol 59. IARC Press, Lyon, France, p 286

IARC 1997 Silica, some silicates, coal dust, and para-aramid fibrils. Vol 68. IARC Press, Lyon, France, p 41–99

IARC 1999 Surgical implants and other foreign bodies. IARC Monographs on the evaluation of carcinogenic risks to humans. Vol 74. IARC Press, Lyon, France, p 311

Isomaki H, Hakulinen T, Joutsenlahti U 1978 Excess risk of lymphomas, leukemia and myeloma in patients with rheumatoid arthritis. J Chronic Dis 31:691

Kaplan RP 1987 Cancer complicating chronic ulcerative and scarifying mucocutaneous disorders. Adv Dermatol 2:19–46

Langman MJS, Cheng KK, Gilman EA, Lancashire RJ 2000 Effect of anti-inflammatory drugs on overall risk of common cancer: case-control study in general practice research database. Br Med J 320:1642–1646

Lowenfels AB, Maisonneuve P, Boyle P, Zatonski WA 1999 Epidemiology of gallbladder cancer. Hepatogastroenterology 46:1529–1532

Mahmoud FA, Rivera NI 2002 The role of C-reactive protein as a prognostic indicator in advanced cancer. Curr Oncol Rep 4:250–255

Mantovani G, Maccio A, Madeddu C et al 2002 Quantitative evaluation of oxidative stress, chronic inflammatory indices, and leptin in cancer patients: correlation with stage and performance status. Int J Cancer 98:84–91

Masferrer JL, Leahy KM, Koki AT et al 2000 Antiangiogenic and antitumor activities of cyclooxygenase-2 inhibitors. Cancer Res 60:1306–1311

Neugut AI, Rosenberg DJ, Ahsan H et al 1998 Association between coronary heart disease and cancers of the breast, prostate and colon. Cancer Epidemiol Biomarkers Prev 7:869–873

Nightingale TE, Gruber J 1994 Helicobacter and human cancer. J Natl Cancer Inst 86: 1505–1509

Paganini-Hill A, Chao A, Ross R, Henderson B 1989 Aspirin use and chronic diseases: a cohort study of the elderly. Br Med J 299:1247–1250

Parsonnet J 1999 Introduction. In: Parsonnet J (ed) Microbes and malignancy — infection as a cause of human cancers. Oxford University Press, New York, p 3–15

Podolsky DK 2002 Inflammatory bowel disease. N Engl J Med 347:417–429

Pradhan AD, Manson JE, Rossouw JE et al 2002 Inflammatory biomarkers, hormone replacement therapy, and incident coronary heart disease: prospective analysis from the Women's Health Initiative observational study. J Am Med Assoc 287:980–987

Robinson WS 1999 Hepatitis B virus and hepatocellular carcinoma. In: Parsonnet J (ed) Microbes and malignancy — infection as a cause of human cancer. Oxford University Press, New York, p 232–288

Ron E 1996 Thyroid cancer. In: Schottenfeld D, Fraumeni JF (eds) Cancer epidemiology and prevention. Oxford University Press, New York, p 1000–1021

Rosenberg L 1995 Nonsteroidal anti-inflammatory drugs and cancer. Prev Med 24:107–109

Rosin MP, Hofseth LJ 1999 Schistosomiasis, bladder and colon cancer. In: Parsonnet J (ed) Microbes and malignancy — infection as a cause of human cancers. Oxford University Press, New York, p 313–345

Schreinemachers D, Everson R 1994 Aspirin use and lung, colon, and breast cancer incidence in a prospective study. Epidemiology 5:138–146

Scotto J, Fears TR, Kraemer KH, Fraumeni JF 1996 Nonmelanoma skin cancer. In: Schottenfeld D, Fraumeni JF (eds) Cancer epidemiology and prevention. Oxford University Press, New York, p 1313–1330

Shacter E, Weitzman SA 2002 Chronic inflammation and cancer. Oncology 16:217–232

Shaheen N, Ransohoff DF 2002 Gastroesophageal reflux, Barrett oesophagus, and esophageal cancer. J Am Med Assoc 287:1982–1986

Sharpe CR, Collett J-P, McNutt M, Belzile E, Boivin J-F, Hanley JA 2000 Nested case-control study of the effects of non-steroidal anti-inflammatory drugs on breast cancer risk and stage. Br J Cancer 83:112–120

Suerbaum S, Michetti P 2002 Helicobacter pylori infection. New Engl J Med 347: 1175–1186

Thamavit W, Shirai T, Ito N 1999 Liver flukes and biliary cancer. In: Parsonnet J (ed) Microbes and malignancy — infection as a cause of human cancers. Oxford University Press, New York, p 346–371

Thun MJ, Namboodiri MM, Calle EE, Flanders WD, Heath CJ Jr 1993 Aspirin use and risk of fatal cancer. Cancer Res 53:1322–1327

Thun MJ, Henley SJ, Patrono C 2002 Nonsteroidal anti-inflammatory drugs as anticancer agents: mechanistic, pharmacologic, and clinical issues. J Natl Cancer Inst 94: 252–266

Wotherspoon AC, Doglioni C, Diss TC et al 1993 Regression of primary low-grade B-cell gastric lymphoma of mucosa-associated lymphoid tissue type after eradication of Helicobacter pylori. Lancet 342:575–577

Yamaguchi N, Kakizoe T 2001 Synergistic interaction between Helicobacter pylori gastritis and diet in gastric cancer. Lancet Oncol 2:88–94

Zaridze D, Borisova E, Maximovitch D, Chkhikvadze V 1999 Aspirin protects against gastric cancer: results of a case-control study from Moscow, Russia. Int J Cancer 82: 473–476

DISCUSSION

Harris: Polymorphisms in any of these pathways could explain why some people are predisposed to cancer by inflammation and infection. Have you any idea what they are from these epidemiology studies?

Thun: At this point the epidemiological studies haven't examined DNA markers of susceptibility to various inflammatory mediators in relation to cancer risk. This is possible because there are now very large prospective studies with thousands of cases, but it hasn't yet been studied.

Mantovani: Are there polymorphisms for *COX-2*, and if so, have they been looked at in terms of susceptibility to neoplasia? This would be the logical connection.

Ristimäki: There are some polymorphisms in the *COX-1* gene which may relate to thrombotic activity or to the response to aspirin, but I do not recall any functional polymorphisms in *COX-2*. When we consider the players of prostanoid biology we also need to consider enzymes upstream and downstream of *COX-2* (PLA2 and the isomerase enzymes, respectively). Also, for every prostanoid there is a separate receptor. In cancer it looks like prostaglandin E2 is the key player, at least in gastrointestinal tract carcinomas. Thus, in addition to prostanoid-forming enzymes one might want to look for polymorphisms of the prostanoid receptor system (for further information see Papafili et al 2002, Lin et al 2002, Humar et al 2000, Spirio et al 1998).

Balkwill: There have been several publications reporting associations with polymorphisms in inflammatory cytokine genes such as *TNF* and *IL1*, and cancer susceptibility and severity, although these have generally been in small numbers of patients (Warzocha et al 1998, Davies et al 2000, Oh et al 2000). In ovarian cancer we now have nearly 600 cases, and we find that a lot of these associations are lost when you look at larger numbers.

Pepper: There are also polymorphisms in *VEGF*, which is the major angiogenic factor. These are found both in the promoter and the 5′ UTR.

Smyth: In inflammatory bowel disease, cancer is quite rare. In those people with inflammatory bowel disease who actually get cancer, has anyone looked to see whether any of their inflammatory markers are different?

Thun: I don't think that particular analysis has been done, but the analysis in Sweden showed that in people with pancolitis from an early age, the risk of developing cancer is actually quite high (40% over 20 year follow-up). The absolute risk depends mostly on the severity of the underlying disease and its duration. Most of the emphasis in ulcerative colitis has been looking at the genetic factors that predispose to this abnormal response to customary bowel flora, rather than being directed specifically at the inflammation.

Rollins: I want to amplify that question a little bit. In thinking about something like ulcerative colitis, it might be instructive to take the opposite view: perhaps the reason some of these patients develop colon cancer is because they have an underlying cancer susceptibility that is somehow related to the same susceptibility that puts them at risk for ulcerative colitis. Inflammation *per se* may just be an epiphenomenon. One of the reasons that I raise this issue is that when we look at the cumulative risk that Michael Thun showed, the patients who develop ulcerative colitis late in life, after age 40, have a more rapid rise in their risk of developing colon cancer. This suggests that inflammation *per se* may not be the causative agent. Perhaps we should be thinking a little sceptically.

Thun: That scepticism is a good thing to have, but let me put it in perspective. What is happening later in life is that the inflammatory process may be promoting an underlying phenomenon that is already more common. With other exposures that predispose to cancer, it is not uncommon that the absolute risk rises faster at older ages, because the exposure accelerates underlying processes that are already underway. Absolute risk is low at younger ages because you are starting *de novo*.

Oppenheim: There's a general observation that the older you are the more cell-mediated (Th1-mediated) the inflammation becomes. Diseases such as mumps and chicken pox are perfect examples: people developing these diseases when they are older experience more of an inflammatory response. Perhaps the over-40 group also has more Th1 reactivity and therefore experiences more inflammation and more COX involvement. This may provide an explanation for this phenomenon of older subjects with colitis developing colon cancer more rapidly.

Harris: Some of the TNF blocking antibody and TNF soluble receptor studies in Crohn's disease and ulcerative colitis imply that TNF could be a key player for these diseases, Crohn's more clearly.

Mantovani: Are we eventually going to get useful information in this respect from patients who are being treated by anti-TNF and anti-IL1 for rheumatoid arthritis?

Feldmann: In the rheumatoid arthritis patients the only cancer risk that is increased is that of lymphomas. It is very striking that the local inflammatory site does not develop cancer, which is a big difference from the bowel. Crohn's disease patients also have an increased lymphoma risk, so there is some common theme. Members of the TNF family are growth factors for precursor cells of lymphoma which form a type of lymphoma.

Balkwill: There is a big experiment going on with all the people who are having anti-TNF therapy, looking at cancer incidence in them. The only study I'm aware of involved following up about 700 patients for 3 years (Day 2002). The incidence of cancer in these people is about what you would expect.

Feldmann: Those studies are flawed by the problems of companies not really wanting to set up proper registers of all patients treated from the beginning. At the moment the consensus evaluation of the data is that there is no overall increased cancer risk, but the judgement is still out concerning lymphomas. Here the problem is that more severe rheumatoid arthritis gets more lymphoma and these are patients getting TNF inhibitors.

Gordon: Michael Thun, you talked about prospective trials, and case control and cohort studies. What would be the best way to look at this?

Thun: There are trials currently underway in patients with Barrett's oesophagus, examining whether selective COX-2 inhibitors can induce regression of dysplasia. There has also been one small trial completed in Linxian, China, testing whether a selective COX-2 inhibitor could induce regression of precursor lesions for squamous cell oesophageal cancer. A high risk of squamous cell cancer of the oesophagus is endemic in this part of China because of nutritional deficiencies. That trial did not show any benefit from treatment with celecoxib with respect to regression of dysplasia. But it wasn't a well-conducted trial. There is a larger trial underway in the USA testing whether celecoxib can induce regression of premalignant adenocarcinoma, and another trial assessing whether celecoxib is effective as adjuvant therapy in patients with Barrett's oesophagus after ablation of the dysplasia. With respect to inflammatory bowel disease there was one epidemiological study from Sweden that showed that sulfasalazine, the salicylate that is given to treat inflammatory bowel disease was associated with lower risk of colorectal cancer. The problem in looking at people who are being given blockers of TNF receptor is that the number of subjects is not large enough to produce adequate results.

Feldmann: The numbers are there now, with over 200 000 on anti-TNF antibody and over 100 000 on TNF receptor Fc.

Rollins: Speaking of numbers, is there any opportunity to mine the data in the Nurses' or Physicians' Health Study looking at salicylate use and cancer incidence?

Thun: Those analyses are included in my paper. The whole problem is that these are observational studies so they don't give randomised evidence of efficacy. The two critical areas in which we currently lack information are the need for randomised evidence of efficacy and for quantitative assessment of safety: what is the balance of the effects of these treatments across a variety of cancers and other endpoints? The large prospective studies can be informative about the second but not about the first.

Feldmann: One of the key things about cancer mutations is that these involve important signalling molecules such as Ras and p53. Is it known whether in the cancers driven by inflammation the incidence of these mutations is the same in the same site of cancer when inflammation is less obvious?

Richmond: Certainly for melanoma, where the early sunburn during childhood is related to later development of cancer, there is a strong correlation between Raf and Ras mutations, as well as p16 mutations in most of those melanoma patients. I don't think the correct study has been done where they can go back and track those patients to examine the early changes prior to the development of the lesions. It would be lovely to have a study like this.

Ristimäki: Colon cancer that is associated with ulcerative colitis is a disease where there is a distinct pattern of genetic changes when you compare it to sporadic colorectal cancer or those appearing in FAP patients. One of these distinct gene alterations is involved with p16. So there are distinct patterns, at least in some adenocarcinomas that arise from inflammatory background when compared with sporadic cases.

Gordon: Let's turn to the infections, which could be easier to analyse, such as *H. pylori* or hepatitis.

Thun: In the case of *H. pylori*, the recommended procedure is to treat the underlying infection. I don't know the natural history of dysplasia in people after successful treatment of *H. pylori*. But characterizing this better would be very useful. In the case of hepatitis, I am not aware that NSAIDs have been used as part of adjuvant treatment. It is something to consider. The only question that will arise concerns whether there will be any adverse effects of giving COX-2 inhibitors to these patients.

Gordon: If we go beyond the COX-2 or arachidonate metabolites, and we don't make any assumptions about the mediators, would this be a good group to utilize?

Thun: I think it is a really important group. Characterizing which inflammatory processes are most important might have huge clinical importance.

Gordon: What do you need? Is it the nature of the cells, or the subtleties of cellular responses, or different kinds of activation of macrophages? How far down can you usefully go?

Thun: This isn't really my area; I'm not sure.

Ristimäki: In certain cases of gastric cancer it is very clear that the *H. pylori* infection has been gone for decades before cancer arises. Thus, the *H. pylori* may make the ground fruitful for carcinogenesis, but the actual cause of this (i.e. bacterial infection) has been long gone. It is the atrophic environment that is prone to genetic changes due to chemical insults. Thus, it may not be always the infection or inflammation itself but rather the histological changes caused by it that makes the individual more susceptible to the neoplastic transformation.

Gordon: One way round this, which we face with chronic diseases such as atherosclerosis, is to use transgenic mice in a controlled way to study spontaneous or chemical carcinogenesis, and then try looking at the inflammatory markers.

Strieter: Let me comment with regard to Th1/Th2 cells in the context of going back and using the tumour itself. The tumour and tumour microenvironment in general in terms of what we might perceive as an inflammatory response is more Th2-like, with a TGFβ predominance. This would allow one to potentially work backwards. The concept would be that if we end up with more of a Th2-like environment, this might be the sort of inflammation that ultimately leads to tumorigenesis and metastasis. In contrast, a Th1 response would eradicate whatever tumour associated antigen would be present and therefore would attenuate the evolution of tumorigenesis.

Harris: I am a little worried about simplistic approaches. Take angiogenesis. The vasculature in every tissue is different. We would expect the mechanisms of angiogenesis to be different in every sort of cancer, although there will be common themes. The same is likely to be true with inflammation. Although there may be some common players, each chronic disease has to be considered in its own right mechanistically.

Gordon: We have come a long way since Virchow in terms of being able to narrow down molecular targets.

Rollins: To go back to the question you posed earlier, I seem to remember that there have been studies giving COX-2 inhibitors to the *Apc/Min* mouse, and that they were effective. So people are beginning to do the kinds of things you are talking about.

Ristimäki: Not only that, but Professor Taketo's group nicely showed that you can delete the *Cox2* gene and see reduction in polyp size and number (allele dependently) in the *Apc* knockout mice (Oshima et al 1996). This genetic deletion is as effective as treating the *Apc* knockout mice with a COX-2 selective drug, suggesting that the drug is indeed attacking through COX-2 enzymes and not via some other targets.

Gordon: What we need is a mouse model that has a high expression of tumours, and then we cross it with a whole range of other knockouts.

Oppenheim: We can think of cancer as a two-signal event: both growth and mutation are needed. Inflammation is a process that stimulates growth and repair, and new cells are being brought in, so the opportunity for mutation is higher. In the models that people have been looking at, where does the gene defect come in? Does the inflammation *per se* also influence the gene defect, or is there some other signal that comes in?

Thun: In the case of the experimental studies related to COX-2, the *Min* mouse inherits one defective *Apc* allele, and has a high rate of losing function in the second allele. In colon cancer generally, the intrinsic processes provide an elevated mutation rate. Studies using the *Min* mouse model look at inhibition or acceleration of the process of tumorigenesis. Another whole line of experimentation in rats involves a chemical carcinogenesis. A known chemical carcinogen is given to rats of various ages. Either at the beginning, after or prior to treatment, NSAID treatment is given, and the tumour occurrence is compared. Many studies of chemically induced colon cancer in rats show that NSAIDs inhibit the development of these cancers. Inhibition can be achieved at a lower dose if NSAID treatment is given before or at the beginning of carcinogen administration. With respect to colon cancer, the commonly used models involve both exogenous initiating agents and endogenous initiation.

Pollard: One of the questions about those sorts of experiments is whether they have been done with genetically disparate mice, where the bone marrow is genetically different from the rest of the animal, for example. This would tell us whether the COX-2 is in immune cells or within some other cell type. Has this been done yet?

Thun: With respect to the knockout mice, where all cells are affected, there is an interesting observation. The *Cox2* knockout mice develop a lower incidence of colorectal cancer, but so do *Cox1* knockout mice. This was not predicted by the COX-2 hypothesis. This observation raises an interesting question as to the role of COX-1, the constitutive form of the enzyme. It is conceivable that COX-1 plays some as yet undefined role in the induction of COX-2. As far as the attempt to figure out in which cells COX-2 is active, I am not aware of studies that have tried to examine this using knockouts in particular tissues. It has been more commonly studied by staining and using biopsy specimens, characterizing where the activity is.

Pollard: It seems to me to be crucial to use, for example a Cre-Lox system or bone marrow transplantation, so that COX-2 isn't in the macrophages.

Gordon: Would you irradiate the animal before the adoptive transfer?

Pollard: It would need a control with normal bone marrow.

Ristimäki: There is one study in which Lewis lung carcinoma cells were injected into COX-2 knockout animals (Williams et al 2000). In this model less angiogenesis and tumour growth was seen in the COX-2 knockout mice than in the wild-type background. This suggests that the stromal cell COX-2 contributes to the behaviour of the tumour cells. Of course, this doesn't dissect out which stromal cells are involved (i.e. vascular endothelial cells, macrophages or fibroblasts).

Pollard: It seems to me that in the context of inflammation and cancer, this is a critical experiment to do.

Ristimäki: It is critical also in the sense that in mice, COX-2 expression is almost exclusively found in the stroma, and not in the epithelial compartment as it is in human tumours. This is a difference between mice and men. Another difference is that all these FAP rodent models are not actually cancer models. They are pre-invasive lesions (adenomas), since the mice die before invasive cancers develop (due to gastrointestinal tract obstruction or bleeding). In humans we want to treat the invasive cancer and not let it metastasize.

Forni: Are there any examples of chronic inflammation that are not linked to an increase of cancer? What about tuberculosis? Is this considered to be a chronic inflammation?

Gordon: The textbooks say chronic inflammations such as tuberculosis are not linked to an increase of cancer, others such as *Schistosomiasis* may be.

Oppenheim: Psoriasis could be an example. It is chronic, inflammatory and long-lasting. I'm not aware that this has been associated with cancer.

Mantovani: It is a strictly Th1 disease.

Ristimäki: What about the other side of the coin? We have organ transplant patients who are treated with immunosuppressive drugs. Do they show a higher prevalence of certain types of cancers?

Gordon: Bob Schreiber has re-investigated the immune surveillance hypothesis in immunodeficient mice and now has evidence that the immune system does play a role in a range of tumour types.

References

Davies FE, Rollinson SJ, Rawstron AC et al 2000 High-producer haplotypes of tumor necrosis factor alpha and lymphotoxin alpha are associated with an increased risk of myeloma and have an improved progression-free survival after treatment. J Clin Oncol 18: 2843–2851

Day R 2002 Adverse reactions to TNF antagonists in rheumatoid arthritis. Lancet 359:540–541

Humar B, Giovanoli O, Wolf A et al 2000 Germline alterations in the cyclooxygenase-2 gene are not associated with the development of extracolonic manifestations in a large swiss familial adenomatous polyposis kindred. Int J Cancer 87:812–817

Lin HJ, Lakkides KM, Keku TO et al 2002 Prostaglandin H synthase 2 variant (Val511Ala) in African Americans may reduce the risk for colorectal neoplasia. Cancer Epidemiol Biomarkers Prev 11:1305–1315

Oh BR, Sasaki M, Perinchery G et al 2000 Frequent genotype changes at −308 and 488 regions of the tumor necrosis factor-α (TNF-alpha) gene in patients with prostate cancer. J Urol 163:1584–1587

Oshima M, Dinchuk JE, Kargman SL et al 1996 Suppression of intestinal polyposis in Apc delta716 knockout mice by inhibition of cyclooxygenase 2 (COX-2). Cell 87:803–809

Papafili A, Hill MR, Brull DJ, McAnulty RJ, Marshall RP, Humphries SE, Laurent GJ 2002 Common promoter variant in cyclooxygenase-2 represses gene expression: evidence of role in acute-phase inflammatory response. Arterioscler Thromb Vasc Biol 22:1631–1636

Spirio LN, Dixon DA, Robertson J et al 1998 The inducible prostaglandin biosynthetic enzyme, cyclooxygenase 2, is not mutated in patients with attenuated adenomatous polyposis coli. Cancer Res 58:4909–4912

Warzocha K, Ribeiro P, Bienvenu J et al 1998 Genetic polymorphisms in the tumor necrosis factor locus influence non-Hodgkin's lymphoma outcome. Blood 91:3574–3581

Williams CS, Tsujii M, Reese J, Dey SK, DuBois RN 2000 Host cyclooxygenase-2 modulates carcinoma growth. J Clin Invest 105:1589–1594

Chemokine-based pathogenetic mechanisms in cancer

Ilaria Conti, Christine Dube and Barrett J. Rollins[1]

Department of Medical Oncology, Dana-Farber Cancer Institute, 44 Binney Street, Boston, MA 02116 and Department of Medicine, Brigham & Women's Hospital, Harvard Medical School, Boston, MA, USA

Abstract. The chemokine system has evolved primarily to control the trafficking of leukocytes during immune or inflammatory responses. However, through their expression of chemokine ligands and receptors, cancers have commandeered various aspects of this host defence system in order to enhance their growth. Although engineered over-expression of some tumour-derived chemokines can stimulate host antitumour responses, this is unlikely to be the reason that tumour cells express them. Rather, a growing body of clinical and laboratory evidence indicates that cancer cells may secrete chemokines in order to attract host cells that supply the tumours with growth and angiogenic factors. In addition, chemokine receptor expression by tumour cells may permit them to use the host's pre-existing leukocyte trafficking system to invade target tissues during metastatic spread. Together, these observations suggest that therapies directed against chemokine ligands or receptors may be beneficial in cancer.

2004 Cancer and inflammation. Wiley, Chichester (Novartis Foundation Symposium 256) p 29–48

The problem

From the very beginning of microscopic tissue analysis, pathologists have appreciated that most human malignancies are accompanied by inflammatory cell infiltration (Virchow 1863). Based on analogies to foreign body and allogeneic transplantation responses, the general assumption has been that these cells are a manifestation of the host's attempt to reject the tumour. And, in some highly contrived experimental settings, elicited inflammatory cells can directly destroy tumours or contribute to their immune recognition.

Leukocytes can be attracted to inflammatory sites in general, and to tumour sites in particular, by a wide variety of bioactive molecules. These include proteins such

---

[1]This paper was presented at the symposium by Barrett J. Rollins to whom all correspondence should be addressed.

as complement or growth factors, small peptides such as formylated tripeptides, and lipids such as leukotrienes. In cancers, however, evidence is accumulating that a substantial proportion of a tumour cell's repertoire of leukocyte chemoattractants is comprised of chemokines. These are low molecular weight (8–15 kDa) proteins that exert their effects by binding to seven-transmembrane spanning G protein-coupled receptors (Rossi & Zlotnik 2000). Although chemokines have effects on epithelial and mesenchymal cells, these effects and their pathophysiological relevance remain poorly characterized. Rather, the most clear-cut biochemical and genetic data confirm their crucial roles in attracting and, occasionally, activating leukocytes. They have been unambiguously implicated in the pathogenesis of several inflammatory diseases and their presumed importance is 'validated' by the chemokine antagonist programmes currently active at most pharmaceutical companies.

Almost any tumour cell can secrete chemokines constitutively or in response to appropriate stimuli. Considering the tumoricidal capacity of activated mononuclear cells, it is certainly possible that the inflammatory cells associated with tumours are on the attack. Is it fair to assume, then, that the tumour-derived chemokines eliciting these cells are participants in host-mediated tumour destruction? Several considerations mitigate against this simple idea. Foremost is the a priori unlikelihood that tumour cells contribute to their own destruction. Cancers are nearly pure Darwinian systems, and the selective pressures for uncontrolled cell growth are sufficiently great that a biologically 'successful' tumour will almost certainly not expend any energy that does not contribute to its growth, much less to its destruction.

That being the case, what is the explanation for tumour cells' secretion of chemokines? One possibility is that they directly contribute to tumour cell proliferation or aggressive behaviour in an autocrine fashion. So far, there is limited evidence that tumour-derived chemokines have cell autonomous effects. Only activation of CXCR4 by its ligand SDF-1 has been rigorously shown to provide anti-apoptotic signals to tumour cells (Zhou et al 2002). Furthermore, no cancer or cancer susceptibility genes have been mapped to chemokine or chemokine receptor loci suggesting that these proteins are not directly involved in evolution of the malignant phenotype. Given what is currently understood about the genetic basis of cancer, this is a very compelling argument that chemokines and their receptors are almost certainly not involved in tumorigenesis per se.

Nonetheless, it remains likely that the chemokine system does play a role in cancer pathobiology for at least two reasons. First, manipulation of chemokines or their receptors affects the behaviour of tumours in a variety of animal models, although it must be admitted that there is, as yet, no validation for this idea in human cancers. Second, the Darwinian argument that tumour cells do not waste metabolic energy can be applied again: it is exceedingly unlikely that a fully transformed cancer cell would bother to synthesize chemokines or their receptors if they were not contributing to tumour 'fitness'.

This brief article will assess the current state of our knowledge about the function of the chemokine system in cancer, with special emphasis on one chemokine in particular, namely MCP-1, or monocyte chemoattractant protein 1. But it will also discuss the potential role other chemokines might play in target organ involvement by metastatic cancers. It will attempt to assign a functional role to chemokines that is consistent with what we currently understand about cancer biology.

Chemokines and their receptors

As noted above, chemokines are low molecular weight proteins that are usually secreted, although there are some examples of membrane-bound chemokines that may or not be cleaved to effect release of a functional chemoattractant (Rossi & Zlotnik 2000). They are best characterized as chemotactic factors for specific leukocyte subsets. The presence of chemokine receptors on non-leukocytic cells implies that this cannot be their only function, but their activities in epithelial cell biology are poorly understood. Over 40 human chemokines have been identified and fortunately the sequenced human genome indicates that there are unlikely to be many more. This diversity appears to have evolved in concert with the evolution of multiple leukocyte types and allows for highly specific control of leukocyte trafficking either to inflammatory foci or as part of basal migratory patterns. Initially, proteins were assigned to the chemokine family based on the presence in those proteins of 2–4 cysteines at conserved positions. However, as structural solutions have been obtained for increasing numbers of chemokines, it has become apparent that they share a high degree of tertiary structural similarity. Despite having disparate target cell and receptor specificities, all chemokines examined to date adopt what has been called the 'chemokine fold' and have nearly superimposable three dimensional structures (Fernandez & Lolis 2002). It is likely that proteomic technology will allow proteins to be classified as chemokines if they are able to assume this characteristic structure.

For now, however, primary structural analysis (i.e. amino acid sequence) permits a helpful classification of chemokines into four subfamilies based on the patterns of amino-acid substitutions around cysteines in the N-terminal domain. Proteins in which the two cysteines in this domain are adjacent belong to the CC subfamily, while those having one amino acid inserted between the cysteine pair belong to the CXC subfamily. Two minor patterns have also been recognized: XC chemokines, of which there are two, have a single cysteine in this domain, and the

TABLE 1   Chemokine Families

Family	Systematic name	Trivial name(s)
CXC	CXCL1	Gro-α, MGSA-α, MIP-2, KC
	CXCL2	Gro-β, MGSA-β, MIP-2α
	CXCL3	Gro-γ, MGSA-γ, MIP-2β
	CXCL4	Platelet factor 4
	CXCL5	ENA-78
	CXCL6	GCP-2
	CXCL7	Platelet basic protein and its products (β-TG, CTAP-III, NAP-2) (human only)
	CXCL8	IL8 (human only)
	CXCL9	Mig
	CXCL10	IP-10
	CXCL11	I-TAC
	CXCL12	SDF-1
	CXCL13	BLC, BCA-1
	CXCL14	MIP-2γ, BRAK, bolekine
	CXCL15	Lungkine (mouse only)
	CXCL16	
CC	CCL1	I-309 (mTCA3)
	CCL2	MCP-1
	CCL3	MIP-1α, LD78
	CCL4	MIP-1β, Act-2
	CCL5	RANTES
	CCL6	c10 (mouse only)
	CCL7	MCP-3, FIC, MARC

sole CX_3C chemokine also known as 'fractalkine' has three amino acids interposed between the cysteine pair. During the natural evolution of the chemokine family, coincident discovery led to the assignment of several different names to the same proteins, many with marginal informational content. A systematic nomenclature has now been introduced and Table 1 provides a list of chemokines indicating both their trivial and systematic names.

Placement of a chemokine into one of the subfamilies does not predict particularly well the type of leukocyte it will attract, although all CXC

TABLE 1 (*Continued*)

Family	Systematic name	Trivial name(s)
CC	CCL8	MCP-2
	CCL9/CCL10	MIP-1γ, MRP-2 (mouse only)
	CCL11	Eotaxin
	CCL12	mMCP-5 (mouse only)
	CCL13	MCP-4, CKβ10 (human only)
	CCL14	HCC-1, CKβ1 (human only)
	CCL15	HCC-2, MIP-5, MIP-1δ (human only)
	CCL16	HCC-4, CKβ12 (human only)
	CCL17	TARC
	CCL18	DC-CK1, PARC, MIP-4, CKβ7 (human only)
	CCL19	MIP-3β, ELC, exodus-3, CKβ11
	CCL20	MIP-3α, LARC, exodus-1, CKβ4
	CCL21	SLC, 6Ckine, exodus-2, TCA4
	CCL22	MDC
	CCL23	MPIF-1, CKβ8, MIP-3 (human only)
	CCL24	MPIF-2, CKβ6, eotaxin-2
	CCL25	TECK, CKβ15
	CCL26	Eotaxin-3, MIP-4α (human only)
	CCL27	CTAK, ESkine, skinkine
	CCL28	MEC (human only)
C	CL1	Lymphotactin, SCM-1α
	CL2	SCM-1β (human only)
CX$_3$C	CX$_3$CL1	Fractalkine

chemokines that have the three amino acid motif ELR (glutamate–leucine–arginine) near their N-termini are potent neutrophil attractants (Clark-Lewis et al 1991, Hebert et al 1991). Rather, subfamily placement corresponds rigorously to the type of receptor a chemokine will activate. So far, all functional chemokine receptors are members of seven transmembrane-spanning, G protein-coupled receptor families. The receptors were named rationally from the start, and at last count 19 had been well characterized. The receptors and their ligand binding specificities are depicted in Fig. 1.

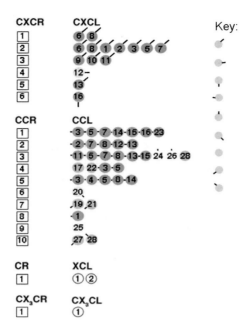

FIG. 1. Chemokine receptor binding specificities. Chemokine receptors are listed by their systematic numerical nomenclature assignment in boxes on the left. Ligands that bind to each receptor with high affinity (generally a $K_d < 10\,\mathrm{nM}$) are shown immediately to the right of each receptor also identified by their systematic numerical assignment. Trivial names for each chemokine can be found by reference to Table 1. The key indicates chromosomal assignment. Chemokines are represented by symbols shown in key to indicate where they cluster at the same locus. (Grateful acknowledgment to Drs Marco Baggiolini and Albert Zlotnik for initially devising this mode of presentation.)

Chemokine specificity

Even a cursory glance at Fig. 1 will reveal one of the central problems in chemokine biology, namely the enormous potential for redundancy. Some receptors, such as CCR3, can have as many as nine high affinity ligands, while some chemokines, such as CCL5 (RANTES) can bind to four different receptors with similar affinities. Most of the chemokines having this high degree of 'redundancy' are induced during inflammatory responses and appear to be responsible for eliciting the influx of circulating leukocytes into tissues. In contrast, chemokines involved in basal leukocyte trafficking, including CXCL13, CCL19, CCL20 and others appear, at the present time, to have single receptor specificities.

While the biology of the constitutively expressed 'non-redundant' chemokines seems to be relatively straightforward, the apparent redundancy of the inducible chemokines poses serious conceptual problems. In addition to the basic biological question they raise (e.g. why should there be so many chemokines having apparently identical functions?) they also present a daunting challenge to pharmacological intervention. There is understandable concern that any attempt to block the activity of a single chemokine or single chemokine receptor involved in inflammation will fail because another member of this family with a similar function will step in and compensate.

Surprisingly, however, there is now strong genetic evidence that this is not the case. Whenever a single chemokine or single chemokine receptor has been genetically inactivated in the mouse, a specific phenotype has resulted. For example, MCP-1/CCL2 has been implicated in atherosclerosis because of its presence in diseased arteries and because of its ability to attract monocytes and macrophages. However, MCP-1/CCL2's receptor, CCR2, binds at least four other ligands with identical affinities and each of these ligands is also a potent monocyte chemoattractant *in vitro* (Rollins 2000). Nonetheless, when disrupted MCP-1/CCL2 alleles are crossed into various atherosclerotic backgrounds, the absence of MCP-1/CCL2 is associated with a 65–85% reduction in aortic lesion formation (Gu et al 1998, Gosling et al 1999). Similarly striking protection has been observed in experimental allergic encephalomyelitis, a rodent model for multiple sclerosis (MS) (Huang et al 2001).

Thus, despite the existence of apparently redundant ligands for CCR2, elimination of only one of these ligands, MCP-1/CCL2, results in an impressive phenotype. The same results have been observed in the analysis of 'knockouts' of several other inducible chemokines and their receptors. One interpretation of these results is that biological specificity among chemokines occurs at the level of expression. That is, while there are perhaps as many as 15 CC chemokines that all attract monocytes with similar potencies, only MCP-1/CCL2 is expressed at high levels in the arterial wall during atherogenesis or in the CNS during exacerbations of MS. The practical consequence is that targeting individual chemokines or their receptors could result in therapeutic benefit.

MCP-1/CCL2 expression in cancer

The concept of chemokine specificity extends to the effects of these proteins in cancer. As noted above, nearly all malignancies are accompanied by a mononuclear cell infiltrate. Even though 15 or more chemokines as well as a number of non-chemokine chemoattractants are all capable of attracting monocytes and macrophages with similar potencies, it appears that a single chemokine, MCP-1/CCL2, is responsible for the majority of tumour cell-derived monocyte chemoattractant activity (Bottazzi et al 1983, Graves et al 1989, Bottazzi et al 1990). This surprising result implies that intervening in this single chemokine system might be therapeutically beneficial. However, the critically important and, as yet, unanswered question is whether MCP-1/CCL2's effects are fundamentally pro- or anti-tumour in nature. Would CCL2/CCR2 blockade help or hurt?

MCP-1/CCL2 can be demonstrated to stimulate host antitumour responses in experimental systems, suggesting that antagonizing MCP-1/CCL2's effects might be counterproductive. For example, malignant Chinese hamster ovary (CHO) cells engineered to express human or murine MCP-1/CCL2 are incapable of forming tumours in nude mice while their non-MCP-1/CCL2-expressing counterparts grow vigorously (Rollins & Sunday 1991). Examination of the tumour implantation site shortly after injection reveals the presence of a robust mononuclear cell infiltrate. Thus MCP-1/CCL2 can elicit an innate, T cell-independent leukocyte response that suppresses tumour formation.

However, MCP-1/CCL2 can also stimulate a T cell-dependent antitumour response. Rat glioma cells transfected with an MCP-1/CCL2 expression vector or a control vector, and subjected to gamma irradiation to prevent their proliferation, have been used to immunize syngeneic, immunocompetent rats (Manome et al 1995). Immunized rats were then challenged by subcutaneous injection of unmanipulated glioma cells. Although immunization using cells transfected with a control vector produced a slight slowing of the challenge tumour's growth rate, the only rats that were completely protected from tumour growth were those that had been immunized with MCP-1-expressing cells. Thus MCP-1/CCL2 can stimulate T cell-dependent as well as T cell-independent antitumour responses.

While these and other demonstrations (Walter et al 1991, Bottazzi et al 1992) of MCP-1/CCL2's antitumour activities are convincing, they take place in highly artificial settings, and merely show that over-expressed chemokines enhance a host's ability to reject tumour xenografts. Furthermore, the Darwinian natural selection argument implies that this is not the reason for tumour cell expression of MCP-1/CCL2 or any other chemokine that can attract and activate a leukocyte. Is it possible, then, that the real reason for chemokine production is to enhance tumour growth? One model consistent with this notion has been championed by Alberto Mantovani and his co-workers (Mantovani et al 1992, Sica et al 2002) who suggest that the tumour elicits a mononuclear infiltrate because these cells provide growth and angiogenic factors that enhance cancer growth.

Clinical epidemiological data support this concept. In a study of 135 women with infiltrating ductal carcinoma of the breast at a variety of stages, MCP-1/CCL2 content of tumour tissue extracts was determined by ELISA (Ueno et al 2000). Patients with high levels of MCP-1/CCL2 in their tumours (defined as greater than the median for the population) had a significantly shorter relapse-free survival than patients with low MCP-1/CCL2 levels. Furthermore, MCP-1/CCL2 expression correlated with the number of tumour infiltrating macrophages in the tumour specimen. That observation is consistent with a large body of literature indicating that macrophage infiltration is an adverse prognostic indicator in breast cancer (Steele et al 1984, Visscher et al 1995, Leek et al 1996).

These results suggest the hypothesis that tumour cells secrete MCP-1/CCL2 in order to attract mononuclear cells that contribute to their growth or metastatic behaviour. One way to test this hypothesis is by examining the behaviour of autochthonous cancers in mice deficient for specific chemokines. For example, we are using $Mcp1/Ccl2^{-/-}$ mice (Lu et al 1998) to analyse the effect of MCP-1/CCL2 on the phenotype of spontaneous mammary cancers driven by transgenic expression of activated rat HER-2/neu in mammary tissue (Boggio et al 1998). Very preliminary results indicate that in the absence of MCP-1/CCL2, tumours grow more slowly and mice survive longer. If confirmed, and if MCP-1/CCL2 expression in this model can be shown to arise in mammary tumour cells, these early findings will support the idea that chemokine expression in endogenous tumours contributes to their growth and biological 'success'. In order to generalize these findings, several additional chemokines must be examined in this and other tumour models.

Chemokines and organ involvement by cancer

It has been suggested that the process whereby circulating tumour cells invade metastatic target organs is broadly analogous to the mechanics of leukocyte emigration (Muller et al 2001). Although the parallels tend to break down on close examination (e.g. tumour cells are generally much larger than leukocytes and may not need adhesion molecule activation for arrest in microvessels), this idea presents a testable hypothesis. A large number of studies have demonstrated that tumour cells express a limited repertoire of chemokine receptors (Soejima & Rollins 2001, Milliken et al 2002), but only a few have provided relevant functional data. Examples include a report that antibodies directed against CXCR4 reduce experimental metastases in a murine breast cancer xenograft model (Muller et al 2001); and melanoma cells engineered to express CCR7 preferentially traffic to regional lymph nodes where the ligands for this receptor are expressed (Wiley et al 2001). Further work will be required to establish whether or not chemokines are generally involved in solid tumour metastases.

Meanwhile, chemokines may play a more easily understandable role in patterns of tissue involvement by cancers derived from cells that normally use chemokine receptors for trafficking (e.g. haematological malignancies). For example, normal basal T cell trafficking is controlled by several different chemokine ligand and receptor pairs depending on the target tissues involved. It now appears that different types of T cell lymphomas express specific chemokine receptors which may account for their characteristic patterns of tissue involvement (Jones et al 2000).

Another haematopoietic cell that is thought to take its migratory cues from chemokines is the Langerhans cell (Dieu et al 1998, Sallusto et al 1998, Sozzani et al 1998). The function of this primary antigen presenting cell of skin is to take up antigen in the local cutaneous environment and transport it to regional lymph nodes where presentation to T cells initiates an immune response. Several lines of evidence indicate that resting Langerhans cells reside in skin because they express the chemokine receptor CCR6 whose ligand, CCL20, is expressed at low levels by basal cutaneous keratinocytes. When the skin's defences are breached by foreign bodies or microorganisms, stimuli such as endotoxin or primary cytokines, e.g. interleukin (IL)1 or tumour necrosis factor (TNF), induce additional local CCL20 production. This attracts Langerhans cells to the inflammatory site where, again under the influence of endotoxin or primary cytokines, they take up antigen and become activated. Part of their activation program is a chemokine receptor switch whereby CCR6 is down-regulated and a new chemokine receptor, CCR7, is up-regulated. The first consequence of this gene expression program is that activated Langerhans cells are no longer anchored in skin because they no longer express the receptor for CCL20. The second consequence is that they are attracted to regional lymph nodes where the CCR7 ligands, CCL19 and CCL21, are synthesized. Travel to the T cell zones of the nodes then provides an opportunity for T-cell activation.

This chemokine-based physiology may be relevant to a malignancy of Langerhans cells called Langerhans cell histiocytosis, or LCH. This is a rare and rather pleiomorphic disease whose variants have been given many different names including histiocytosis X, Letterer–Siwe disease, eosinophilic granuloma and others. However, all of these disorders are united by the fact that the abnormal histiocyte in each one is derived from a Langerhans cell. These diseases are characterized by an overabundance of abnormal histiocytes that involve a variety of tissue sites, most commonly secondary lymphoid organs, skin and bone.

One possible explanation for the pattern of tissue involvement in LCH might be found in the repertoire of chemokine receptors expressed by abnormal Langerhans cells. In a recent study of 24 cases of LCH, immunohistochemical analysis revealed coincident expression of CCR6 and CCR7 by abnormal Langerhans cells in all cases (Fleming et al 2003). Thus these partially activated cells fail to down-regulate CCR6 when they up-regulate CCR7. Furthermore, CCL20 expression was documented both in involved skin and bone. These findings suggest a model in which coexpression of CCR6 and CCR7 directs abnormal Langerhans cells to skin and bone in response to CCL20, and to secondary lymphoid organs in response to CCL19 and CCL21. Similar patterns of chemokine receptor expression have been observed in other histiocytoses that have the same sort of tissue involvement as

LCH (Fleming et al 2003). Antagonism of CCR6 or CCR7 could, at the very least, promote emigration of these activated cells from involved organs thereby leading to symptomatic relief. In addition, however, since some chemokines have been shown to provide survival signals for malignant cells (Zhou et al 2002), interrupting CCR6 or CCR7 signalling might also lead to apoptosis of abnormal Langerhans cells.

Summary

The chemokine system evolved *pari passu* with functionally specialized leukocytes; as the latter's number and complexity increased, so did the former. This has provided an opportunity for enterprising parasites, such as malignant cells, to commandeer the migratory control system for their own ends. Examples cited above include the recruitment of mononuclear cells for the growth and angiogenic stimuli they provide, and the cooptation of trafficking mechanisms that permit the establishment of metastatic outposts in anatomically disparate and favourable environments. It is now up to us to be as creative as tumour cells in developing the means to disrupt their use of this complex inflammatory system.

Acknowledgments

Supported by a grant from the University of Pisa to IC; USPHS grants CA53091 and AI50225, and a generous grant from Team Histio of the Dana-Farber Cancer Institute/Jimmy Fund Marathon Walk to BJR.

References

Boggio K, Nicoletti G, Di Carlo E et al 1998 Interleukin 12-mediated prevention of spontaneous mammary adenocarcinomas in two lines of Her-2/neu transgenic mice. J Exp Med 188:589–596

Bottazzi B, Polentarutti N, Acero R et al 1983 Regulation of the macrophage content of neoplasms by chemoattractant. Science 220:210–212

Bottazzi B, Colotta F, Sica A, Nobili N, Mantovani A 1990 A chemoattractant expressed in human sarcoma cells (tumor-derived chemotactic factor, TDCF) is identical to monocyte chemoattractant protein-1/monocyte chemotactic and activating factor (MCP-1/MCAF). Int J Cancer 45:795–797

Bottazzi B, Walter S, Govoni D, Colotta F, Mantovani A 1992 Monocyte chemotactic cytokine gene transfer modulates macrophage infiltration, growth, and susceptibility to IL-2 therapy of a murine melanoma. J Immunol 148:1280–1285

Clark-Lewis I, Schumacher C, Baggiolini M, Moser B 1991 Structure-activity relationships of interleukin-8 determined using chemically synthesized analogs: critical role of NH_2-terminal residues and evidence for uncoupling of neutrophil chemotaxis, exocytosis, and receptor binding activities. J Biol Chem 266:23128–23134

Dieu MC, Vanbervliet B, Vicari A et al 1998 Selective recruitment of immature and mature dendritic cells by distinct chemokines expressed in different anatomic sites. J Exp Med 188:373–386

Fernandez EJ, Lolis E 2002 Structure, function, and inhibition of chemokines. Annu Rev Pharmacol Toxicol 42:469–499

Fleming MD, Pinkus JL, Alexander SW et al 2003 Coincident expression of the chemokine receptors CCR6 and CCR7 by pathologic Langerhans cells in Langerhans cell histiocytosis. Blood 101:2473–2475

Gosling J, Slaymaker S, Gu L et al 1999 MCP-1 deficiency reduces susceptibility to atherosclerosis in mice that overexpress human apolipoprotein B. J Clin Invest 103:773–778

Graves DT, Jiang YL, Williamson MJ, Valente AJ 1989 Identification of monocyte chemotactic activity produced by malignant cells. Science 245:1490–1493

Gu L, Okada Y, Clinton SK et al 1998 Absence of monocyte chemoattractant protein-1 reduces atherosclerosis in low density lipoprotein receptor-deficient mice. Mol Cell 2:275–281

Hebert CA, Vitangcol RV, Baker JB 1991 Scanning mutagenesis of interleukin-8 identifies a cluster of residues required for receptor binding. J Biol Chem 266:18989–18994

Huang DR, Wang J, Kivisakk P, Rollins BJ, Ransohoff RM 2001 Absence of monocyte chemoattractant protein 1 in mice leads to decreased local macrophage recruitment and antigen-specific T helper cell type 1 immune response in experimental autoimmune encephalomyelitis. J Exp Med 193:713–726

Jones D, O'Hara C, Kraus MD et al 2000 Expression pattern of T-cell-associated chemokine receptors and their chemokines correlates with specific subtypes of T-cell non-Hodgkin lymphoma. Blood 96:685–690

Leek RD, Lewis CE, Whitehouse R, Greenall M, Clarke J, Harris AL 1996 Association of macrophage infiltration with angiogenesis and prognosis in invasive breast carcinoma. Cancer Res 56:4625–4629

Lu B, Rutledge BJ, Gu L et al 1998 Abnormalities in monocyte recruitment and cytokine expression in monocyte chemoattractant protein 1-deficient mice. J Exp Med 187: 601–608

Manome Y, Wen PY, Hershowitz A et al 1995 Monocyte chemoattractant protein-1 (MCP-1) gene transduction: an effective tumor vaccine strategy for non-intracranial tumors. Cancer Immunol Immunother 41:227–235

Mantovani A, Bottazzi B, Colotta F, Sozzani S, Ruco L 1992 The origin and function of tumor-associated macrophages. Immunol Today 13:265–270

Milliken D, Scotton C, Raju S, Balkwill F, Wilson J 2002 Analysis of chemokines and chemokine receptor expression in ovarian cancer ascites. Clin Cancer Res 8:1108–1114

Muller A, Homey B, Soto H et al 2001 Involvement of chemokine receptors in breast cancer metastasis. Nature 410:50–56

Rollins BJ 2000 MCP-1, -2, -3, -4, -5. In: Oppenheim JJ, Feldman S, Durum S, Hirano T, Vilcek J, Nicola N (eds) Cytokine reference. Academic Press, London, p 1145–1160

Rollins BJ, Sunday ME 1991 Suppression of tumor formation *in vivo* by expression of the *JE* gene in malignant cells. Mol Cell Biol 11:3125–3131

Rossi D, Zlotnik A 2000 The biology of chemokines and their receptors. Annu Rev Immunol 18:217–242

Sallusto F, Schaerli P, Loetscher P et al 1998 Rapid and coordinated switch in chemokine receptor expression during dendritic cell maturation. Eur J Immunol 28:2760–2769

Sica A, Saccani A, Mantovani A 2002 Tumor-associated macrophages: a molecular perspective. Int Immunopharmacol 2:1045–1054

Soejima K, Rollins BJ 2001 A functional IFN-gamma-inducible protein-10/CXCL10-specific receptor expressed by epithelial and endothelial cells that is neither CXCR3 nor glycosaminoglycan. J Immunol 167:6576–6582

Sozzani S, Allavena P, D'Amico G et al 1998 Differential regulation of chemokine receptors during dendritic cell maturation: a model for their trafficking properties. J Immunol 161:1083–1086

Steele R J, Eremin O, Brown M, Hawkins RA 1984 A high macrophage content in human breast
    cancer is not associated with favourable prognostic factors. Br J Surg 71:456–458
Ueno T, Toi M, Saji H et al 2000 Significance of macrophage chemoattractant protein-1 in
    macrophage recruitment, angiogenesis, and survival in human breast cancer. Clin Cancer
    Res 6:3282–3289
Virchow R 1863 Die krankhaften Geschwulste. Dreißig Vorlesungen, gehalten während des
    Wintersemesters 1862–1863 an der Universität zu Berlin. In: Vorlesungen uber Pathologie.
    Verlag von August Hirschwald, Berlin
Visscher DW, Tabaczka P, Long D, Crissman JD 1995 Clinicopathologic analysis of
    macrophage infiltrates in breast carcinoma. Pathol Res Pract 191:1133–1139
Walter S, Bottazzi B, Govoni D, Colotta F, Mantovani A 1991 Macrophage infiltration and
    growth of sarcoma clones expressing different amounts of monocyte chemotactic protein/
    JE. Int J Cancer 49:431–435
Wiley HE, Gonzalez EB, Maki W, Wu MT, Hwang ST 2001 Expression of CC chemokine
    receptor-7 and regional lymph node metastasis of B16 murine melanoma. J Natl Cancer Inst
    93:1638–1643
Zhou Y, Larsen PH, Hao C, Yong VW 2002 CXCR4 is a major chemokine receptor on glioma
    cells and mediates their survival. J Biol Chem 277:49481–49487

DISCUSSION

Balkwill: We have heard a very MCP-1-centred paper. Is it the opinion of people who work in this field that MCP-1 is a very important chemokine in cancer, or is it just the one that we study? For instance, there is quite a lot of RANTES in breast cancer, and these data are quite compelling.

Rollins: A while ago Dana Graves and Tony Valente looked at 30–40 tumour cell lines, all of which expressed monocyte chemoattractant activity. When that activity was purified the majority of it was all due to MCP-1. Alberto Mantovani has seen the same thing, essentially, that the primary monocyte chemoattractant made by most tumour cells is MCP-1. I am not going to answer your question directly because I don't know for sure that MCP-1 is an important chemokine in cancer. But it is clearly the chemokine that is primarily responsible for attracting mononuclear cells to the site. Why is it made? For the same reason that other NF-κB responsive genes are transcribed in tumour cells.

Blankenstein: I always thought that CSF-1 had similar biological functions.

Gordon: It is actually not identical to MCP-1.

Pollard: There are many chemoattractants for macrophages. CSF-1 is an extremely potent chemoattractant, and there is quite a lot of clinical evidence in certain tumours that MCP-1 is made at high levels. In fact, in our mouse models, we think that CSF-1 is not a particularly important chemoattractant.

Blankenstein: Then our conclusion is that you need both.

Pollard: CSF-1 is different, because it is a growth factor for mononuclear phagocytes. It is the required lineage growth factor for mononuclear phagocytes. It is a pluripotent growth factor with a role in survival, differentiation, proliferation and chemotaxis. It is also chemokinetic for macrophages. It is a multipotent pleiotrophic growth factor, as opposed to a rather specific chemoattractant.

Wilkins: If there is a lot of redundancy in the system of production and specificity of action, then you would expect that there would be few of these genes that would map as oncogenes. In contrast, you would expect many of them to act as tumour suppressors when they are mutated to loss of function. The results you reported suggest this. Has any one done a rigorous search in tumour-resistant genotypes in humans or mice to see whether any of these genes correlate as quantitative trait loci?

Rollins: They haven't shown up, to my knowledge.

Balkwill: Going back to MCP-1, we mustn't just think about the chemokines; we have to also consider the receptors. Certainly, our experience in ovarian cancer is that MCP-1 is the predominant chemokine, but when we look for the receptor we don't find CCR2 in the tumour. It is down-regulated. I think there are many chemokines in the tumours but not many receptors. This may be the key to the specificity of the system.

Gordon: The receptors might be on the leukocytes.

Balkwill: No, we have looked. They are not even found on those cells.

Rollins: Is CCR2 down-regulated by the ligands?

Balkwill: Yes, but when we look in ascites of ovarian cancer we find very high levels of the chemokine in the fluid, and we can find the receptor on tumour cells and leucocytes in ascites.

Balkwill: Is the infiltrate any different in the tumours that you get in *Mcp1* knockout mice?

Rollins: It is hard to say from simply looking at Haemotoxylin & Eosin sections. We haven't yet stained for specific cell phenotypes in those lesions.

Strieter: Did you look at levels of MCP-1/CCL2 protein expression? What were the quantitative differences between the heterozygotes and the wild-type in those two models?

Rollins: We haven't been able to look in the tumours yet. If we look at peritoneal macrophages from heterozygous mouse there is about half the normal amount of MCP-1 in the heterozygotes. There is a haploinsufficient state.

Brennan: Are they compromised in being able to recruit macrophages?

Rollins: We never looked at the heterozygotes when we did our challenges.

Blankenstein: I also have a question concerning the gene dosage effect. In your experiments with crossing *op/op* and *Her2*-transgenic mice did you compare the original mice from Guido Forni to your F1 backcross?

Rollins: No, the controls here were siblings of the heterozygous mice. We are very careful about making sure that when we looked at homozygous and heterozygous MCP-1 mice that they came from the same parents. Allele assortment can be a huge problem.

Mantovani: The dissociation between primary tumour appearance and growth rate, and metastasis, is reminiscent of something we observed long ago when we did gene transfer. We had a dissociation between primary tumours and metastasis. This may relate to the relative number of macrophages. In other words, disseminating tumour cells encounter higher numbers of phagocytes than those present in the primary lesions.

Gordon: Could it be that the cells are doing both protection of the animal and promotion of the tumour at the same time? Different subpopulations are found in different places. It is a mixed story.

Oppenheim: The presumption is that in this situation the macrophages that are attracted are serving as a source of trophic factors that is promoting tumour growth, and that by interfering with tumour infiltration by macrophages we will interfere with these trophic effects. On the other hand, Alberto Mantovani has shown that macrophages can also have antitumour effects and promote tumour regression in fibrosarcoma. In a collaborative study with him, we found that MCP-1 was promoting the influx of macrophages that were interfering with the fibrosarcoma. I remember that you presented papers about biphasic effects of MCP-1, and that it can have opposing effects depending on the model. This makes the situation very complicated. It is hard to apply Darwinian selection theory to the idea that MCP-1 at some times can help tumour growth and at other times can interfere. We have had situations in which antibody to MCP-1 could interfere with tumour growth. This surprised us, because of the literature suggesting that it could actually have the opposite effects. How do we resolve all this? Is there any circumstance under which we could predict which way MCP-1 will influence the tumour?

Mantovani: I believe that under physiopathological conditions the prevailing function of CCL2/MCP-1 is the pro-tumour function. Tying together all the evidence, unless you over-express the gene and get over-recruitment, the physiological mechanisms of production and recruitment are pro-tumour.

Blankenstein: Tumour transplantation experiments and primary tumour models are difficult to compare with each other. One should not exclude the possibility that macrophages are anti-tumorigenic in tumour transplantation models, but pro-tumorigenic in primary tumour models. The situation is even more problematic, since I will present an example that macrophages can also be pro-tumorigenic in tumour transplantation experiments.

Pollard: That is exactly the point I was going to make. We need to be careful about defining models here. Xenotransplant experiments probably have very little to do with tumorigenesis. One should really be thinking about the spontaneously occurring or oncogene-driven tumours, where there is a normal microenvironment and probably not lots of transplantation antigens being presented. This may yield a very different result. Is there any experiment where the removal of MCP-1 has had a negative impact on a naturally occurring tumour model, such as this ErbB2-generated model?

Rollins: It has all been transplantation experiments.

Pollard: Maybe we are actually looking at transplant rejection here, and not tumour rejection. This makes the issue a little less confusing.

Forni: Cytokines also influence the progression of these tumours. Injection of exogenous cytokines such as IL2 and IL12 in the HER-2/neu transgenic mice significantly impairs tumour progression. There is always a two-stage effect, stimulation versus inhibition, which probably depends on the amount of cytokine.

Blankenstein: The experiment with the *Mcp1* knockouts is quite similar to the one that you did with the *Csf1* knockout mice. If I remember correctly, it is not the initiation but the progression of the tumour that is impaired in the *Csf1* knockout mice. Do you have any indication that this is similar in *Mcp1* knockout mice?

Rollins: I think that is an accurate picture. As a first approximation from what we can see, the incidence of these multifocal tumours that arise in the HER-2/neu model is not different in the heterozygous knockout mice. Again, we need to do a truly mechanistic study, but my guess at this point is that we are affecting the progression phase rather than the initiation, if only because activated HER-2/neu is such a powerful oncogene, I seriously doubt whether manipulating the macrophage environment is going to affect the expression and effects of HER-2/neu overexpression.

Pollard: That is exactly the situation for us. We use the polyoma middle T oncogene, which is such a powerful oncogene that 100% of the mice have tumours by 6 weeks of age. We saw no effect on incidence, but we are trying to stop a sledgehammer.

Gordon: Are you studying a spontaneous tumour model?

Pollard: This is in an MMTV polyoma middle T driven model, and we are not planning at present to use a spontaneous tumour model.

Ristimäki: It is very important to have models like this where we actually get an invasive metastasizing cancer, because these are very rare in rodents. In clinics we don't really worry about the primary tumour progression, which we can remove. It is the metastasis that kills the patients. This is the mechanism that should be studied more carefully.

Pollard: As you say, many of the rodent models don't metastasize. There are very few that do, and of these some are very peculiar. We have worked with the MMTV-Wnt1 model, which we have sort of rejected now, because the whole mammary epithelium comes down with a hyperplasia and we only occasionally get adenocarcinomas. The only time they are ever invasive tumours is if there is some surgery of the mammary gland, and then when the mammary glands are allowed to recover they get invasive tumours. At least with the polyoma middle T model, which we have now documented very carefully, we start with a single primary tumour with essentially unaffected epithelium all round it. This then progresses through various stages and becomes metastatic. Later on we see secondary tumours on the distal ducts, but they are several stages of progression behind the primary tumour.

Richmond: I wanted to challenge this Darwinian argument suggesting that there is no genetic evidence that chemokine expression leads to cell autonomy. Is it possible that in your final argument with CCR6 and CCR7 co-expression in the LCH model you have an event which leads to continuous expression of two chemokine receptors that does lead to relative cell autonomy. The cells begin to proliferate and expand in that compartment.

Rollins: I don't think that those changes lead to cell autonomy. The most conservative interpretation of that result would be that it leads to patterns of tissue involvement, and that there is some other event that can be genetically linked that leads to actual cell autonomy. These pathological Langerhans cells have p53 abnormalities and a variety of other abnormalities. They probably represent a stage of arrested Langerhans cell development, similar to many other cancers. Whatever is the cause of that is the event that leads to cell autonomous behaviour. I'd make a bet right now that when an LCH gene is mapped, it is not going to map to CCR6 or CCR7, but to a controller of phenotypic differentiation or a transcription factor. It will fit the paradigm of other cancers that we have seen. I have yet to encounter a cancer gene that is related directly to chemokines.

Richmond: Barrett Rollins, when do you think the p53 mutations occur in the LCH model? Before CCR6/CCR7-expressing cells become activated? Or is it possible that the p53 mutations occur as the result of being in an environment where you have accumulated a group of mononuclear cells that are activated and proliferating?

Rollins: No one knows.

Oppenheim: I have a physiological question. Does the simultaneous expression of CCR6 and CCR7 represent a normal stage?

Caux: In normal CD34$^+$ haematopoietic progenitor cell differentiation you can see concomitant expression of both CCR6 and CCR7. But this is a transient stage.

D'Incalci: I was confused by some aspects of your paper. I wanted to understand whether MCP-1 had a positive or negative effect. You said there are about 40 or 50 chemokines, and so I imagine that the system is very complex and would require an evaluation of many factors rather than just looking at one. Therefore, it is not so surprising that the different models show different things, because there is probably an interplay of all these factors. I wonder whether, as in other fields, it is possible to look at patterns of chemokines rather than just one chemokine. You have a nice model, but is it really relevant to look at just one factor?

Rollins: I apologise for confusing you, but I blame the field also, because it is a confusing one. What you are saying is correct, but the amount of information one can derive from a reductionist approach can be very large. In the chemokine field in particular there is this myth that because there are so many chemokines and receptors and because there is so much overlap in the specificity of certain receptors for certain ligands, that the system is replete with redundancy, and focusing on a single chemokine is pursuing a false hope. In fact, I think every single knockout of a chemokine or chemokine receptor has resulted in a phenotype, which tells you that at a physiological level there is less redundancy than one might have predicted. Having done all this reductionist sort of work, one has to admit that at the end of the day it is necessary to try to reintegrate this with what we know about all the other chemokines or cytokines that are expressed in a particular setting. But I would still argue that there is a lot to be gained from at least initially beginning with reductionist approaches.

Gordon: Can I defend immunology by quoting Henry James who said there is no problem with being both positive and negative at the same time.

Mantovani: I have a question about the second part of Barrett Rollins' paper. If we look in the circulation at the levels of MIP-3α, can they serve as a marker for the disease? This would be very useful.

Rollins: We haven't looked at serum levels of MIP-3α.

Strieter: In terms of the latter part of your talk, your immunohistochemistry suggested that there is less CCR6 than CCR7, although this is clearly qualitative. Is there any evidence to suggest that one receptor might be preferential over the other in terms of function?

Rollins: As you say, we can't infer anything quantitative from the immunohistochemistry. These were two completely different antibodies. We are now trying to make pathological Langerhans cell lines by putting telomerase in, but it is very hard to obtain fresh tissues because it is a rare disease.

Brennan: You suggested that a CCR6 antagonist might induce the migration away from the skin to the lymph nodes. Have you any way of testing this? I wasn't clear whether there were any mouse models of these particular types of colon cancers.

Rollins: The only way I can think of testing this is by using antibodies. The CCR6 knockouts are highly immunodeficient and have no secondary lymphoid structures to begin with. We can't really use the knockouts in any informative way. Antibodies against the murine receptor would be one way of approaching this.

Brennan: When you mention the CCR6 antagonist, do you mean an antibody or a chemokine that blocks?

Rollins: Anything that antagonizes signalling through the receptor.

Oppenheim: If I remember correctly there are data that some tumours can express low levels of receptors for chemokines, and respond to low level chemokine production by various tissues. This may then lead to their selective metastasis to particular organs. We had a model in which a T cell tumour variant was able to go to the kidney in its metastatic spread, whereas the parental line went elsewhere but not the kidney. It turns out that the receptor that was expressed by the variant was functional, but the one in the parent wasn't. Similar data have been reported by Zlotnik and co-workers (Muller et al 2001) indicating that expression of the receptors on a melanoma tumour cell line can lead to the spread to certain tissues under certain circumstances. There is a mirror-image role of receptors being able to have effects in terms of directing how tumours may invade and where they may go.

Gordon: Since you have raised this, many of these chemokine receptors are not great adhesion molecules, but compared with the other adhesion molecules surely they play a much greater role in terms of tissue localization, spread, retention and so on.

Oppenheim: Chemokines regulate adhesion molecules and immediately reconfigure them.

Gordon: So you are saying their effect is indirect rather than direct.

Oppenheim: Yes. They are like the general telling the sergeants what to do. Chemokines control the adhesion process very well.

Gordon: Who controls the chemokines?

Oppenheim: Everybody else! In other words, the cytokines and exogenous stimulants.

Balkwill: The experiments I like the best in terms of the tumour chemokine receptors are those of Sam Hwang and his colleagues (Wiley et al 2001, Murakami et al 2002). If they overexpress CCR7 in B16 melanoma cells they can stimulate lymph node metastases, but if they overexpress CXCR4 in the same cells they are 'posted' to the lungs.

Hermans: There might be another layer of subtlety to this in terms of the dosage of chemokines or receptors involved. Does a 15% reduction in MCP-1 attract a different type of monocyte, for example? Has anyone studied this kind of dosage effect?

Rollins: That's an interesting idea, but I can't think of any studies addressing this.

Mantovani: The only thing that comes to mind is that in humans there is a subset of monocytes that is CD16 positive. It was reported that these cells have higher levels of CCR2 and respond to lower doses of CCL2/MCP-1. These are more efficient at least in terms of inflammatory cytokine production. In a way, you could somehow skew them by dose.

References

Muller A, Homey B, Soto H et al 2001 Involvement of chemokine receptors in breast cancer metastasis. Nature 410:50–56

Murakami T, Maki W, Cardones AR et al 2002 Expression of CXC chemokine receptor-4 enhances the pulmonary metastatic potential of murine B16 melanoma cells Cancer Res 62:7328–7334

Wiley HE, Gonzalez EB, Maki S, Wu M-T, Hwang ST 2001 Expression of CC chemokine receptor-7 and regional lymph node metastasis of B16 murine melanoma. J Natl Cancer Inst 93:1638–1643

General discussion I

Smyth: I have a very simple question that stems from the first two papers we have seen. Barrett Rollins showed some data on human breast cancers where growth rate was increased. Is it possible that the final common pathway of all interactions between inflammatory modulators and cancer promotion could be reduced to a kinetic basis, of a hyperkinetic state where there is more cell division and therefore greater opportunity for epigenetic change and growth promotion? In inflammatory bowel disease, one of the consequences is that there is much greater turnover of normal colonic crypt cells because of the response to inflammation. The data for MCP-1 and breast cancer are quite persuasive: growth rate is increased.

Rollins: I think that is an excellent way to look at the progression side of things. If you increase growth rate, this increases the target for genetic abnormalities. In addition to simply increasing proliferative rate, in order for this to happen *in vivo*, for these tumours to proliferate beyond a certain size it is necessary to activate angiogenic pathways. The infiltrate may be providing not only growth enhancers directly to the tumour cell, it may also be enhancing that growth through epiphenomena such as creating more tumour-specific blood vessels.

Smyth: Do pro-angiogenic factors alter the kinetics of tumour growth?

Harris: Yes. If you transfect them into breast cancer cell lines they do increase growth.

Richmond: Are there data out there suggesting that constitutive expression of chemokines is pro-tumorigenic? I agree with Barrett Rollins that independent genetic mutations probably aren't occurring in the chemokines which are then linked to cancer, but in the transgenic model where we developed mice over-expressing the MIP-2 chemokine in melanocytes using the tyrosinase promoter/enhancer to drive the *Mip2* transgene expression in mice lacking p16, DMBA treatment clearly resulted in an increased incidence of the development of melanoma in the C57BL background, which incidentally is very resistant to the development of melanoma lesions. Consequently, I think that over-expression of pro-tumorigenic chemokines combined with genetic alterations may be involved in tumour progression. We all believe that it takes at least two hits to cause a tumour. Over-expression of chemokine can in fact be a factor which enhances tumour progression.

Oppenheim: One of the few things chemokines don't do is to be directly mitogenic for endothelial cells. They are differentiative and promote survival but they aren't mitogenic. On the other hand, through the promotion of angiogenic activities, for example, they can be very supportive of tumour growth. Interestingly enough, many of them that are proinflammatory are also proangiogenic.

Feldmann: It may not be well known to all the audience that the chronic inflammatory diseases have a very strong pro-angiogenic component. This is certainly true for rheumatoid arthritis (RA), Crohn's disease and atherosclerosis. In our own work in RA what we have found is that the expression of at least one angiogenic factor, vascular endothelial growth factor (VEGF), precedes very large synovial mass. In an early arthritis clinic, where we get a variety of different arthritis patients, some of whom will have rheumatoid arthritis and some of whom have other conditions, we did a prospective study. We measured serum VEGF, which is measuring a mediator in the wrong place, but which is a reflection of total VEGF body synthesis. We found that all the early RA patients had already expressed VEGF before we could make a formal diagnosis. It may well be that we have to look for a whole range of parallels: mitogenic, chemotactic and angiogenic stimuli. It may be that there is a pattern of gene expression which we need to address.

Ristimäki: Going back to the cytokine story there used to be some paradoxical cytokines that *in vivo* were protumorigenic but *in vitro* actually inhibited tumour cell growth. Angiogenesis is clearly part of the story explaining why they work *in vivo* in promotion of tumorigenicity. Clinically, when a pathologist gets a section of inflamed bowel, even if there is no tumour the epithelial cells can look quite neoplastic. This disappears when we heal the inflammation. There may be some signal directed at the epithelial cells pushing them to put on programs that make them think that they are actually cancer cells. After healing this is shut off. This is why it is not a neoplasia but just a reactive change. Understanding this shut off mechanism should prove to be useful.

Pepper: We work on lymphangiogenesis and we, like others in the field, suggest that this may be a new therapeutic opportunity in cancer. Which patients are we targeting? Are we targeting the primary tumour or the process of metastasis or the growth of the secondary tumour? We have heard already that it is the secondary tumour that will kill the patient. What is the chemokine profile of the secondary tumour versus the primary tumour? If we are looking at chemokines in the primary tumour, is this a reflection of what we should be treating? I accept this may be an impossible question to answer, but we should keep in mind who and what we are targeting when we are talking about inhibitors.

Strieter: It is important to consider that primary tumours have the potential to metastasize. The mechanisms of this potential are targetable. We therefore reduce metastasis by targeting the primary tumour in this regard. By the same token we have opportunities because not all malignant cells are the same, and those that have metastatic potential may be markedly different. In the context of our model systems, we look at the primary tumour, and in particular interleukin (IL)8 expression that drives angiogenesis in this model. 35% of the tumour cells are expressing this (these are A549 adenocarcinoma cell lines), yet when we look at the metastatic lesion over 85% of them are expressing this molecule. You could target the primary tumour in terms of IL8 which may then lead to targeting both the metastatic lesion as well as the primary tumour. There are specific issues that could be targeted that don't necessarily affect the primary tumour size or angiogenesis. I will show data in my paper that you could target a molecule that has nothing to do but specifically attenuate metastatic potential. There are multiple targets.

Ristimäki: In breast cancer we have a big problem. We want to minimize the treatment in case the tumour is small, in which case most patients do not need additional treatment. You don't want to treat a patient who has a low possibility of having metastasis already present. Still these are the very patients for which we should identify markers that would predict poor prognosis. We need the markers that tell us whether a patient needs this harsh treatment or not. We have some already, but we need more. Perhaps these chemokines can be of help.

Pollard: In the microarray-type experiments have there been any patterns of chemokine expression that correlate with clinical prognosis?

Balkwill: This hasn't been done yet, but there is no reason why it couldn't be.

Harris: There have been three or four breast cancer arrays published now with 8000 genes each. Chemokines didn't come up strongly. In the melanoma array *WNT5* came up.

Strieter: I would argue that this is a flaw of the microarray analyses.

Pollard: 8000 genes is a relatively small set, and we would have to ask how many chemokine genes are actually on the array.

Rollins: The current Affymetrix chips only cover about 70% of the known chemokines and chemokine receptors.

Harris: Those that we have talked about, such as MCP, MCF1 and TNF will be on those.

Strieter: The proteins have longer half-life than the mRNA that is expressed.

Pollard: They might also represent a small component of the microenvironment.

Brennan: That is a point. When they did the microarrays on rheumatoid tissue, TNF was always absent.

Balkwill: That is true of some tumours as well. You don't high levels of these cytokines and chemokines. For instance, CCL5 (RANTES) is expressed by a very small proportion of the stromal cells in ovarian cancer. There was a significant correlation between that small amount of RANTES production and the extent of the infiltrate in a series of primary tumours. Sometimes very small amounts of cytokines and chemokines can have a profound effect that we don't understand.

Ristimäki: I have a technical point about the chips. I think that it would be a good idea to try to do the arrays separately on the stromal and the neoplastic epithelial compartment of the tumours.

Anti-TNFα therapy of rheumatoid arthritis: what can we learn about chronic disease?

Marc Feldmann, Fionula M. Brennan, Ewa Paleolog, Andrew Cope, Peter Taylor, Richard Williams, Jim Woody* and Ravinder N. Maini

*Kennedy Institute of Rheumatology Division, Faculty of Medicine, Imperial College of Science, Technology and Medicine, 1 Aspenlea Road, London W6 8LH, UK and *Roche Bioscience, 3401 Hillview Avenue, Palo Alto, CA 94304, USA*

Abstract. The importance of tumour necrosis factor (TNF)α in rheumatoid arthritis (RA) was initially proposed on the basis of analysis of cytokine gene regulation at the local site of the disease, the synovium. This was then verified in animal models and established in an extensive series of clinical trials, culminating in now 250 000 treated patients with either of two approved TNF inhibitors, antibody or fusion protein. The degree and magnitude of clinical benefit has enabled analyses of the mechanism by which anti-TNF benefits, and hence insights into important steps in the disease process. It was found that essentially all aspects of RA were ameliorated, and important mechanisms of benefit involved diminution of multiple pro-inflammatory cytokines, adhesion molecules and chemokines, leading to reduced cell trafficking, reduced angiogenesis and most importantly halting of joint destruction. What of the problems? Safety is better than prior drugs, but there is a small increase in severe infections, smaller than might have been anticipated. Cost is the major drawback limiting greater use. In view of the central pathological processes down-regulated, and their role in many diseases, the early clinical success of anti-TNF in RA led to subsequent successful trials and registration in Crohn's disease and juvenile rheumatoid arthritis, and successful trials in ankylosing spondylitis, psoriasis and psoriatic arthritis. The era of anti-cytokine therapeutics is just dawning.

2004 Cancer and inflammation. Wiley, Chichester (Novartis Foundation Symposium 256) p 53–73

Rheumatoid arthritis (RA) is one of the commonest chronic causes of disability (Panayi et al 2001). It is an autoimmune/inflammatory disease of complex and poorly understood aetiology. The only clear cut genetic basis is the MHC with the shared epitope of the HLA-DR4/DR1 β chain (QKARRA etc) at positions 71–75, but other possible relevant genes include the tumour necrosis factor (TNF)α gene (Gregerson et al 1987). In identical twins, concordance is only 15–30%, indicating the importance of extrinsic, non-inherited aspects (Silman et al

53

1993). These remain to be identified, but smoking is one (Silman et al 1996), as are microbial agents (Holoshitz et al 1986, Ueno et al 1993, Takahashi et al 1998, Kingsley 1997), the role of which, long suspected, remains to be established (Albani et al 1995).

In contrast to the paucity of information concerning aetiology, our understanding of the pathogenesis continues to be enriched by a variety of clinical studies. In this chapter we will review the current state of understanding of the role of proinflammatory cytokines, especially in this disease and its therapy.

Current status and how we got there

Rationale

Multiple groups were interested in the expression of proinflammatory cytokines in RA tissue in the 1980s, as the technology to assay them accurately was developed consequent to cloning cytokine cDNA. A plethora of proinflammatory cytokines were described in these tissues, either in synovial fluid (SF), synovial membrane (SM), or short-term cultures of SF or SM cells (see for example Xu et al 1989, Di Giovine et al 1988, Feldmann et al 1996a). A representation of these results is shown in Table 1.

Essentially all conceivable cytokines are present, with the exception of interleukin (IL)4. This lack may be relevant in terms of the described Th1 skewing in RA, and the importance of IL4 in the generation of Th2 cells. But a key problem was to understand how this plethora of cytokines was involved in the disease process. Were any or all therapeutic targets?

Our approach to resolve this issue was to study the regulation of cytokines within the synovial tissue. We used short term cultures of entire synovial membrane, which are complex mixtures of chiefly T lymphocytes and macrophages, with dendritic cells, B lymphocytes, endothelial cells and so on. They produce a mixture of proinflammatory cytokines, over the 5 or 6 days of culture employed without any extrinsic stimulation (Brennan et al 1989a). Since IL1 had been described in the 1980s by Saklatvala et al (1985) as a major inducer of joint damage *in vitro* or in animal models, we began our studies with the regulation of IL1.

In principle there were many powerful stimuli that might regulate IL1 in synovium, such as immune complexes, interferon (IFN)γ, IL1 itself, TNFα, lymphotoxin and GM-CSF. However, we found a dramatic effect of anti-TNFα anti-serum on IL1 production in RA synovial cultures, but not osteoarthritis synovial cultures (Brennan et al 1989b). This provided the first clue that TNFα might have a special role amongst proinflammatory cytokines, data that were mirrored in mice given Gram-negative bacteria, where TNFα is the earliest

TABLE 1   Cytokine expression in rheumatoid arthritis

Cytokine	Expression	
	mRNA	protein
Pro-inflammatory		
IL1α and IL1β (interleukin 1)	+¶	+
TNFα	+	+
LT (lymphotoxin)	+	±
IL6	+	+
GM-CSF (granulocyte macrophage colony stimulating factor)	+	+
M-CSF (macrophage colony stimulating factor)	+	+
LIF (leukocyte inhibitory factor)	+	+
Oncostatin M	+	+
IL2	+	±
IL3	−	−
IL7	?	?
IL9	?	?
IL12	+	+
IL15	+	+
IFNα, β (interferon α/β)	+	+
IFNγ	+	±
IL17	+	+
IL18	+	+
Immunoregulatory		
IL4	±	−
IL10	+	+
IL11	+	+
IL13	+	+
TGFβ (transforming growth factor β)	+	+
Chemokines		
IL8	+	+
Groα (melanoma growth stimulating activity)	+	+
MIP-1 (macrophage inflammatory protein)	+	+
MCP-1 (monocyte chemoattractant protein)	+	+
ENA-78 (epithelial neutrophil activating peptide 78)	+	+
RANTES (regulated upon activation T cell expressed & secreted)	+	+
Mitogens		
FGF (fibroblast growth factor)	+	+
PDGF (platelet-derived growth factor)	+	+
VEGF (vascular endothelial growth factor)	+	+

¶Cytokines expressed in rheumatoid synovial tissue: +, present; −, absent. Modified after Feldmann et al (1996a).

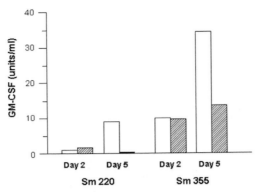

FIG. 1.   GM-CSF production by RA synovial cell cultures, inhibited with neutralising antibodies to TNFα. RA synovial joint cells were cultured for periods up to 5 days with (hatched bars) or without (open bars) neutralizing polyclonal antibodies to TNFα (Reproduced with permission from Haworth et al 1991).

response cytokine, and anti-TNFα blocked the subsequent production of IL1 and IL6 (Fong et al 1989).

Follow-on studies showed anti-TNFα blocked synovial production of other proinflammatory cytokines, including GM-CSF, IL8 and IL6 (Haworth et al 1991, Butler et al 1995), raising the concept of a TNFα-dependent cytokine network (Feldmann et al 1996a) and suggesting that TNFα might be a possible therapeutic target (Brennan et al 1989b, Feldmann et al 1990, Brennan et al 1992) (Fig. 1). The conceptual problem of the potential futility of blocking just one of a large set of proinflammatory cytokines with similar properties, leaving the others present to drive the disease process was resolved. In rheumatoid synovial cultures *in vitro*, TNFα blockade down-regulated a cluster of other relevant proinflammatory cytokines (Fig. 2).

The potential utility of this concept was further validated in the animal model of RA, collagen-induced arthritis, where hamster anti-murine TNF was injected at high dose (12 mg/kg) after disease onset, and successfully ameliorated both inflammation and joint destruction, results almost simultaneously obtained in three laboratories (Thorbecke et al 1992, Williams et al 1992, Piguet et al 1992) (Fig. 3).

These results in animal models paved the way for the first clinical trial of anti-TNF in RA patients, performed in 1992 at the Kennedy Institute/Charing Cross Hospital. All 20 patients improved to a readily detectable extent (Elliott et al 1993), and thus anti-cytokine therapy was on its way (Fig. 4). Formal double blind randomised placebo-controlled trials followed (Fig. 5) (Elliott et al 1994), which included detailed mechanism of action studies (Paleolog et al 1996, 1998, Tak et al 1996, Charles et al 1999, Ulfgren et al 2000).

TIMP-1, TIMP-2

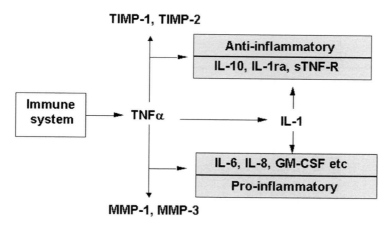

FIG. 2. Cytokine cascade in RA. Pro- and anti-inflammatory cytokines interact in a 'network' or 'cascade' (Reproduced with permission from Feldmann et al 1996b).

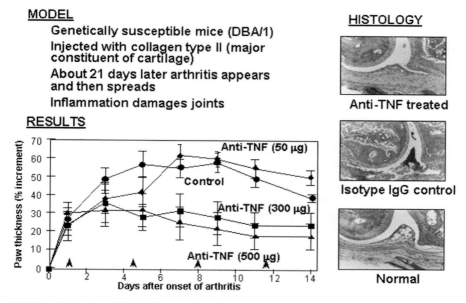

FIG. 3. Animal model of collagen-induced arthritis. The graph shows the effects of different doses of anti-TNF on clinical progression of established arthritis. Arrows indicate time of injections of anti-TNF. Paw width was measured using calipers and increase in thickness expressed as a percentage compared with baseline. Histology: paraffin sections of paws were stained with haematoxylin. Bottom: normal joint; middle: severe arthritis; top: mouse treated with anti-TNF (Reproduced with permission from Williams et al 1992).

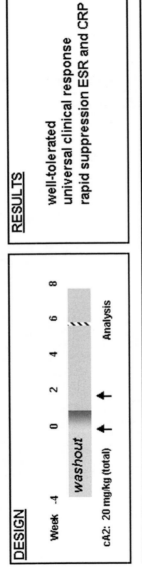
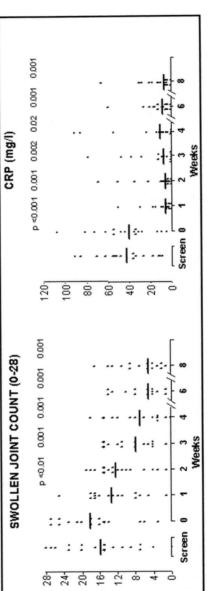

FIG. 4. Open-label treatment with infliximab in RA (Reproduced with permission from Elliott et al 1993).

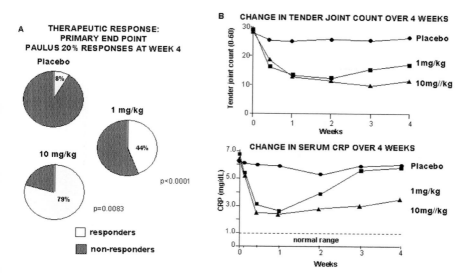

FIG. 5. Formal proof of efficacy: randomised, double-blind placebo controlled trial of infliximab. In a randomized double-blind trial patients had a single infusion of infliximab at 1 mg/kg or 10 mg/kg body weight or placebo, on day 0. (A) Changes in Paulus 20% criteria at 4 weeks. (B) Changes in tender joint counts and serum C-reactive protein levels. (Reproduced with permission from Elliott et al 1994).

Mechanism of action

The marked clinical benefit in the majority of treated patients permitted mechanism of action studies, which illuminate the pathogenesis of RA. Essentially all aspects of the disease are influenced towards normality. An important aspect of mechanism is reduction in recruitment of leucocytes (Fig. 6) (Taylor et al 2000).

Clinical use

Currently two anti-TNF agents are in routine use for RA. The first to be approved was a dimeric TNF-R p75 IgG-Fc fusion protein; this is given subcutaneously, 25 mg twice per week, and is known as etanercept or Enbrel® (Moreland et al 1999, 1997).

Infliximab, the anti-TNF chimeric (mouse Fv, human IgG1) antibody which we used, known as Remicade® (Knight et al 1993), is infused intravenously every 8 weeks, at 3 mg/kg, in combination with methotrexate. The latter combination treatment is based on the success of phase II trial, where this combination was markedly synergistic at low anti-TNF dose (1 mg/kg) and synergistic at later time

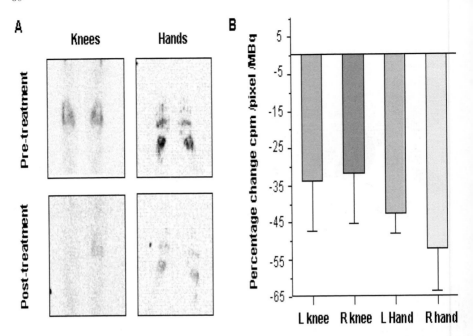

FIG. 6.   Infliximab monotherapy reduces retention of granulocytes in rheumatoid joints. (A) Anterior gamma-camera images of the knees and hands of a patient with rheumatoid arthritis taken over 20 minutes at 22 hours after bolus injection of autologous radiolabelled granulocytes, before and after a single 10 mg/kg intravenous bolus of infliximab. (B) Percentage of change from baseline in the injected dose of ^{111}In-labelled granulocytes at 2 weeks after a single 10 mg/kg intravenous bolus of infliximab. Data were expressed as the geometric mean cpm per 100 pixels divided by the injected dose of radioactivity (in MBq) (Reproduced with permission from Taylor et al 2000).

points at 3 or 10 mg/kg (Fig. 7). Methotrexate has a complex, not fully understood mode of action, which includes anti-T cell effects, promoting apoptosis and reducing IFNγ production, and thus was thought capable of mimicking the anti-TNF/anti-CD4 synergy so potent in animal models (Williams et al 1994) (Fig. 8). The clinical results bear this out, and most importantly have demonstrated consistently over many trials and routine clinical use that methotrexate reduces the immunogenicity of infliximab (Fig. 9), and in its absence infusion reactions and anti-idiotype antibodies were produced at much greater frequency (Feldmann & Maini 2001, Maini & Taylor 2000).

Current numbers of patients which have been treated with TNF inhibitors are not easy to establish, but sales to date (September 2002) indicate that over 300 000 patients have been treated with either etanercept or infliximab, despite a shortage of etanercept. With the expected launch of adalimumab (D2E7) next year clinical use

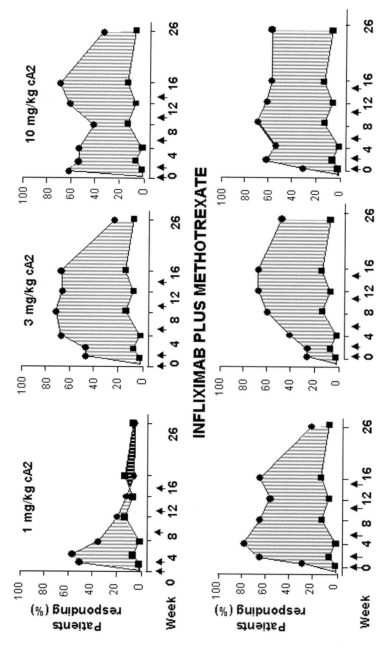

FIG. 7. Methotrexate (MTX) enhances clinical response to infliximab. The percentage of patients responding according to Paulus 20% criteria at weeks 1, 2, 4, 8, 16 and 26 in a double-blind, placebo-controlled trial of 101 rheumatoid arthritis patients treated without or with methotrexate at 7.5 mg/week and infliximab as indicated. Arrows indicate times of infliximab infusions. ■, placebo with/without methotrexate; ●, infliximab with/without methotrexate. (*Upper panels*) Monotherapy (without MTX). (*Lower panels*) Combination therapy (MTX plus infliximab) (Reproduced with permission from Maini et al 1998).

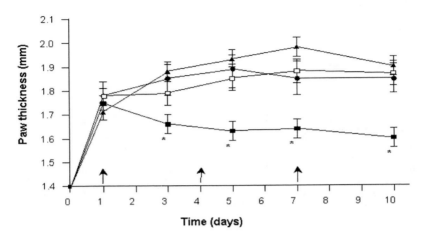

FIG. 8.   Combination therapy with sub-optimal anti-TNF plus anti-CD4 in collagen-induced arthritis. Comparison of the effect of an optimal dose of anti-TNFα (300 μg) alone versus the same dose of anti-TNFα plus anti-CD4 (200 μg) on paw thickness in murine collagen-induced arthritis. Arrows indicate times of injection. Paw width was measured using calipers. An asterisk indicates a significant reduction to the group given control antibody (*P* < 0.05). (□) anti-TNFα alone; (●) anti-CD4 alone; (■) anti-TNFα plus anti-CD4; (▲) control antibody (Reproduced with permission from Williams et al 1994).

is likely to climb further (Rau 2002). Even so, it is used in a very small percentage of rheumatoid patients globally.

Joint protection

The advent of TNF inhibiting biologicals has provided not only benefit in terms of symptoms, and signs, but also improvements in quality of life and markedly reduced or halted rate of joint destruction, as judged by serial X-rays (Lipsky et al 2000a,b) (Fig. 10).

The correlation between quality of life, joint destruction and prognosis is not well documented as yet, but it is difficult to believe that halting joint damage will not be reflected in an improved prognosis. Whether reduction in C-reactive protein (CRP), fibrinogen and platelet levels will also result in a reduction in the elevated cardiovascular mortality in RA remains to be documented.

Problems with anti-TNF therapy

These include cost, convenience and side effects.

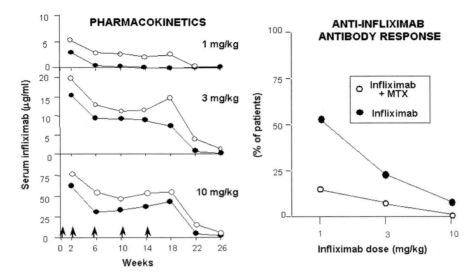

FIG. 9.   The influence of concomitant methotrexate on pharmacokinetics of infliximab. (*Left panel*) Patients receiving infliximab at 1 mg/kg, 3 mg/kg and 10 mg/kg with concomitant methotrexate (○) and without methotrexate (●). Blood was drawn immediately prior to infliximab infusion up to 14 weeks and further blood samples were taken every 2 weeks thereafter through to week 26 (excluding week 24). Infusions of infliximab at 0, 2, 6, 10 and 14 weeks. (*Right panel*) Incidence of anti-infliximab antibodies at each dose of infliximab with (○) and without methotrexate (●) (Reproduced with permission from Maini et al 1998).

Cost

At $12–15 000 per year, etanercept and infliximab are expensive. However with their efficacy in recalcitrant patients, it was concluded by the National Institute for Clinical Excellence (NICE) that these were cost-effective and should be prescribed by the NHS (NHS National Institute for Clinical Excellence 2002).

It is likely that marked reduction in cost will only occur after much more competition emerges, particularly from small molecular chemical medicines. What these will be is unclear, but p38 MAP kinase or IKK-2 inhibitors appear to be the most promising candidates under development in multiple pharmaceutical companies (Woronicz et al 1997, Li et al 1999, Lee et al 2000, Badger et al 2000).

Side effects

Rheumatoid patients, even without anti-TNF therapy have a higher risk of infection than comparable adults (Maini et al 1999). Active RA inhibits the

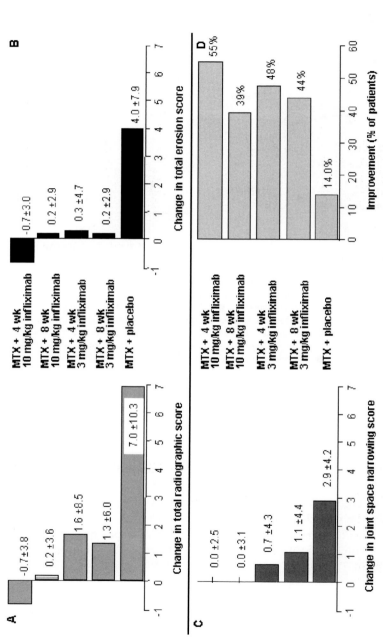

FIG. 10. Effects of infliximab and methotrexate on radiographs of hands and feet demonstrating joint protection at 52 weeks in the ATTRACT trial. Joint damage was assessed radiographically using the van der Heijde modification of the Sharp scoring system; the higher the score the greater the articular damage. Measurements in placebo or infliximab treated groups, all receiving concomitant MTX, are shown. (*Panel A*) Change from baseline in total radiographic score (mean ± SD). (*Panel B*) Change from baseline in total erosion scores. (*Panel C*) Change from baseline in total joint space narrowing scores. (*Panel D*) Percentage improvement of patients whose radiographs showed a reduction (i.e. minus values) in total radiographic scores from baseline (Based on Lipsky et al 2000a).

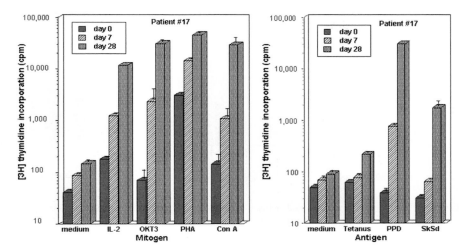

FIG. 11. Chronic exposure to tumor necrosis factor (TNF) *in vitro* impairs the activation of T cells through the T cell receptor/CD3 complex; reversal *in vivo* by anti-TNF antibodies in patients with rheumatoid arthritis (Reproduced with permission from Cope et al 1994).

immune response, especially at the T-cell level, and this appears to be due to TNFα as it thus can be induced rapidly *in vitro* (Cope et al 1992). The diminished T cell response can be restored by anti-TNF therapy *in vivo* (Cope et al 1994) (Fig. 11).

This may be the reason that the risk of infection after anti-TNF therapy is not high, not nearly as high as animal model studies of the role of TNFα in protection from infection would predict (Keane et al 2001).

The future

It seems clear that the benefit due to TNF blockade happens locally in the synovium, while side effects are due to TNF blockade elsewhere. So is it conceivable to block TNFα in synovium, but leave TNF function elsewhere? There are different possible approaches. One is to physically deliver TNF inhibitors to the joint. The obvious approach, to inject into the joint space, as is done for steroids, is unlikely to work; there is no evidence that antibody will penetrate synovium, unlike the cell permeable corticosteroids, which have intracellular receptors. Might it be possible to selectively target activated cells, more prevalent in joints? There are technologies which may be able to do that. But perhaps the most interesting is to target differences between TNFα synthesis in RA synovium and in the immune system. This has been shown to be possible by our recent studies with the TNFα produced by macrophages in joints being driven

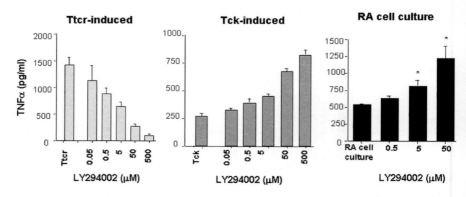

FIG. 12.   Evidence that rheumatoid arthritis synovial T cells are similar to cytokine-activated T cells (Reproduced with permission from Brennan et al 2002).

in large part by cytokine-activated T cells (Brennan et al 2002) (Fig. 12). Blocking these cells would provide synovial specific TNFα inhibition, and thus appears to be a promising therapeutic strategy.

Conclusions

Concepts about pathogenesis of human disease need verification by clinical trials to be fully accepted. It is now abundantly clear that pro-inflammatory cytokines are of major importance in the pathogenesis of RA, with the role of TNFα being validated by its clinical benefit in hundreds of thousands of patients. But there is still a lot to learn, as there are no cures. The impetus for further research in this field is clear. Our techniques for understanding pathogenesis and establishing new therapies are effective, and this will lead to improvements in the future.

References

Albani S, Keystone EC, Nelson JL et al 1995 Positive selection in autoimmunity: abnormal immune responses to a bacterial DNAJ antigenic determinant in patients with early rheumatoid arthritis. Nat Med 1:448–152

Badger AM, Griswold DE, Kapadia R et al 2000 Disease-modifying activity of SB 242235, a selective inhibitor of p38 mitogen-activated protein kinase, in rat adjuvant-induced arthritis. Arthritis Rheum 43:175–183

Brennan FM, Chantry D, Jackson AM, Maini RN, Feldmann M 1989a Cytokine production in culture by cells isolated from the synovial membrane. J Autoimmun 2:S177–S186

Brennan FM, Chantry D, Jackson A, Maini R, Feldmann M 1989b Inhibitory effect of TNF alpha antibodies on synovial cell interleukin-1 production in rheumatoid arthritis. Lancet 2:244–247

Brennan FM, Maini RN, Feldmann M 1992 TNFα — a pivotal role in rheumatoid arthritis? Br J Rheumatol 31:293–298

Brennan FM, Hayes AL, Ciesielski CJ, Green P, Foxwell BMJ, Feldmann M 2002 Evidence that rheumatoid arthritis synovial T cells are similar to cytokine-activated T cells: involvement of phosphatidylinositol 3-kinase and nuclear factor kappaB pathways in tumor necrosis factor alpha production in rheumatoid arthritis. Arthritis Rheum 46:31–41

Butler DM, Maini RN, Feldmann M, Brennan FM 1995 Modulation of proinflammatory cytokine release in rheumatoid synovial membrane cell cultures. Comparison of monoclonal anti-TNFalpha antibody with the interleukin-1 receptor antagonist. Eur Cytokine Network 6:225–230

Charles P, Elliott MJ, Davis D et al 1999 Regulation of cytokines, cytokine inhibitors, and acute-phase proteins following anti-TNF-alpha therapy in rheumatoid arthritis. J Immunol 163:1521–1528

Cope AP, Aderka D, Doherty M et al 1992 Increased levels of soluble tumor necrosis factor receptors in the sera and synovial fluid of patients with rheumatic diseases. Arthritis Rheum 35:1160–1169

Cope AP, Londei M, Chu NR et al 1994 Chronic exposure to tumor necrosis factor (TNF) in vitro impairs the activation of T cells through the T cell receptor/CD3 complex; reversal in vivo by anti-TNF antibodies in patients with rheumatoid arthritis. J Clin Invest 94:749–760

Di Giovine FS, Meager A, Leung H, Duff GW 1988 Immunoreactive tumour necrosis factor alpha and biological inhibitor(s) in synovial fluids from rheumatic patients. Int J Immunopathol Pharmacol 1:17–26

Elliott MJ, Maini RN, Feldmann M et al 1993 Treatment of rheumatoid arthritis with chimeric monoclonal antibodies to tumor necrosis factor alpha. Arthritis Rheum 36:1681–1690

Elliott MJ, Maini RN, Feldmann M et al 1994 Randomised double-blind comparison of chimeric monoclonal antibody to tumour necrosis factor alpha (cA2) versus placebo in rheumatoid arthritis. Lancet 344:1105–1110

Feldmann M, Maini RN 2001 Anti-TNFα therapy or rheumatoid arthritis: What have we learned? Annu Rev Immunol 19:163–196

Feldmann M, Brennan FM, Chantry D et al 1990 Cytokine production in the rheumatoid joint: implications for treatment. Ann Rheum Dis 49:480–486

Feldmann M, Brennan FM, Maini RN 1996a Role of cytokines in rheumatoid arthritis. Annu Rev Immunol 14:397–440

Feldmann M, Brennan FM, Maini RN 1996b Rheumatoid arthritis. Cell 85:307–310

Fong Y, Tracey KJ, Moldawer LL et al 1989 Antibodies to cachectin/tumor necrosis factor reduce interleukin 1beta and interleukin 6 appearance during lethal bacteremia. J Exp Med 170:1627–1633

Gregersen PK, Silver J, Winchester RJ 1987 The shared epitope hypothesis. An approach to understanding the molecular genetics of susceptibility to rheumatoid arthritis. Arthritis Rheum 30:1205–1213

Haworth C, Brennan FM, Chantry D, Turner M, Maini RN, Feldmann M 1991 Expression of granulocyte-macrophage colony-stimulating factor in rheumatoid arthritis: regulation by tumor necrosis factor-alpha. Eur J Immunol 21:2575–2579

Holoshitz J, Klajman A, Drucker I et al 1986 T lymphocytes of rheumatoid arthritis patients show augmented reactivity to a fraction of mycobacteria cross-reactive with cartilage. Lancet 2:305–309

Keane J, Gershon S, Wise RP et al 2001 Tuberculosis associated with infliximab, a tumor necrosis factor alpha-neutralizing agent. N Engl J Med 345:1098–1104

Kingsley G 1997 Microbial DNA in the synovium — a role in aetiology or a mere bystander? Lancet 349:1038–1039

Knight DM, Trinh H, Le J et al 1993 Construction and initial characterization of a mouse-human chimeric anti-TNF antibody. Mol Immunol 30:1443–1453

Lee JC, Kumar S, Griswold DE, Underwood DC, Votta BJ, Adams JL 2000 Inhibition of p38 MAP kinase as a therapeutic strategy. Immunopharmacology 47:185–201

Li ZW, Chu W, Hu Y et al 1999 The IKKbeta subunit of IkappaB kinase (IKK) is essential for nuclear factor kappaB activation and prevention of apoptosis. J Exp Med 189:1839–1845

Lipsky PE, van der Heijde DMFM, St Clair EW et al 2000a Infliximab and methotrexate in the treatment of rheumatoid arthritis. Anti-Tumor Necrosis Factor Trial in Rheumatoid Arthritis with Concomitant Therapy Study Group. New Engl J Med 343: 1594–1602

Lipsky PE, van der Heijde DMFM, St Clair EW et al 2000b 102 week clinical and radiological results from the ATTRACT trial: A 2 year, randomized, controlled, phase 3 trial of infliximab (Remicade) in patients with active rheumatoid arthritis despite methotrexate. Arthritis Rheum 47:S242

Maini RN, Taylor PC 2000 Anti-cytokine therapy for rheumatoid arthritis. Annu Rev Med 51:207–229

Maini RN, Breedveld FC, Kalden JR et al 1998 Therapeutic efficacy of multiple intravenous infusions of anti-tumor necrosis factor alpha monoclonal antibody combined with low-dose weekly methotrexate in rheumatoid arthritis. Arthritis Rheum 41:1552–1563

Maini R, St Clair EW, Breedveld F et al 1999 Infliximab (chimeric anti-tumour necrosis factor alpha monoclonal antibody) versus placebo in rheumatoid arthritis patients receiving concomitant methotrexate: a randomised phase III trial. ATTRACT Study Group. Lancet 354:1932–1939

Moreland LW, Baumgartner SW, Schiff MH et al 1997 Treatment of rheumatoid arthritis with a recombinant human tumor necrosis factor receptor (p75)-Fc fusion protein. N Engl J Med 337:141–147

Moreland LW, Schiff MH, Baumgartner SW et al 1999 Etanercept therapy in rheumatoid arthritis. A randomized controlled trial. Ann Intern Med 130:478–486

NHS National Institute for Clinical Excellence 2002 Guidance on the use of etanercept and infliximab for the treatment of rheumatoid arthritis. Technical Appraisal Guidance No.36

Paleolog EM, Hunt M, Elliott MJ, Feldmann M, Maini RN, Woody JN 1996 Deactivation of vascular endothelium by monoclonal anti-tumor necrosis factor alpha antibody in rheumatoid arthritis. Arthritis Rheum 39:1082–1091

Paleolog EM, Young S, Stark AC, McCloskey RV, Feldmann M, Maini RN 1998 Modulation of angiogenic vascular endothelial growth factor by tumor necrosis factor alpha and interleukin-1 in rheumatoid arthritis. Arthritis Rheum 41:1258–1265

Panayi GS, Corrigall VM, Pitzalis C 2001 Pathogenesis of rheumatoid arthritis. The role of T cells and other beasts. Rheum Dis Clin North Am 27:317–334

Piguet PF, Grau GE, Vesin C, Loetscher H, Gentz R, Lesslauer W 1992 Evolution of collagen arthritis in mice is arrested by treatment with anti-tumour necrosis factor (TNF) antibody or a recombinant soluble TNF receptor. Immunology 77:510–514

Rau R 2002 Adalimumab (a fully human anti-tumour necrosis factor alpha monoclonal antibody) in the treatment of active rheumatoid arthritis: the initial results of five trials. Ann Rheum Dis 61:II70–II73

Saklatvala J, Sarsfield SJ, Townsend Y 1985 Pig interleukin 1. Purification of two immunologically different leukocyte proteins that cause cartilage resorption, lymphocyte activation, and fever. J Exp Med 162:1208–1222

Silman AJ, MacGregor AJ, Thomson W et al 1993 Twin concordance rates for rheumatoid arthritis: results from a nationwide study. Br J Rheumatol 32:903–907

Silman AJ, Newman J, MacGregor AJ 1996 Cigarette smoking increases the risk of rheumatoid arthritis. Results from a nationwide study of disease-discordant twins. Arthritis Rheum 39:732–735

Tak PP, Taylor PC, Breedveld FC et al 1996 Decrease in cellularity and expression of adhesion molecules by anti-tumor necrosis factor a monoclonal antibody treatment in patients with rheumatoid arthritis. Arthritis Rheum 39:1077–1081

Takahashi Y, Murai C, Shibata S et al 1998 Human parvovirus B19 as a causative agent for rheumatoid arthritis. Proc Natl Acad Sci USA 95:8227–8232

Taylor PC, Peters AM, Paleolog E et al 2000 Reduction of chemokine levels and leukocyte traffic to joints by tumor necrosis factor alpha blockade in patients with rheumatoid arthritis. Arthritis Rheum 43:38–47

Thorbecke G J, Shah R, Leu CH, Kuruvilla AP, Hardison AM, Palladino MA 1992 Involvement of endogenous tumor necrosis factor alpha and transforming growth factor beta during induction of collagen type II arthritis in mice. Proc Natl Acad Sci USA 89:7375–7379

Ueno Y, Umadome H, Shimodera M, Kishimoto I, Ikegaya K, Yamauchi T 1993 Human parvovirus B19 and arthritis. Lancet 341:1280

Ulfgren AK, Andersson U, Engstrom M, Klareskog L, Maini RN, Taylor PC 2000 Systemic anti-tumor necrosis factor alpha therapy in rheumatoid arthritis down-regulates synovial tumor necrosis factor alpha synthesis. Arthritis Rheum 43:2391–2396

Williams RO, Feldmann M, Maini RN 1992 Anti-tumor necrosis factor ameliorates joint disease in murine collagen-induced arthritis. Proc Natl Acad Sci USA 89:9784–9788

Williams RO, Mason L J, Feldmann M, Maini RN 1994 Synergy between anti-CD4 and anti-tumor necrosis factor in the amelioration of established collagen-induced arthritis. Proc Natl Acad Sci USA 91:2762–2766

Woronicz JD, Gao X, Cao Z, Rothe M, Goeddel DV 1997 IkappaB kinase-beta: NF-kappaB activation and complex formation with IkappaB kinase-alpha and NIK. Science 278:866–869

Xu WD, Firestein GS, Taetle R, Kaushansky K, Zvaifler NJ 1989 Cytokines in chronic inflammatory arthritis. II. Granulocyte-macrophage colony-stimulating factor in rheumatoid synovial effusions. J Clin Invest 83:876–882

DISCUSSION

Thun: I have several questions related to the potential for studying the populations treated with anti-TNF therapy in epidemiological studies. Of the 300 000 or so patients who are being treated, what fraction are being treated for Crohn's disease as opposed to rheumatoid arthritis. How long has that treatment been used? What is the probable age distribution of the people being treated?

Feldmann: The age distribution of patients with rheumatoid and Crohn's is different, with the Crohn's patients being a little younger. The approval for anti-TNF treatment in Crohn's is just three doses of infliximab (Remicade) and then you have to stop. You can imagine what this does to creative US physicians. This is not a situation where follow up will happen, because the rules make this difficult. Trials of chronic use are underway. My understanding is that the number of Crohn's patients treated is roughly 60 000.

Thun: I agree with what you said earlier, that this would be a really interesting population to study. It would be great if the pharmaceutical companies could somehow facilitate this. Presently, the barriers you mentioned will make it tough. At least in the USA the only way we could get at this would be through something like Medicare, in which case we would be picking up people aged over 65. It does seem like there is an opportunity being missed here.

Feldmann: It is amazing what opportunities are missed. One of the most interesting things is that the risk factors such as fibrinogen, platelets and CRP for cardiovascular disease are dramatically changed by anti-TNF therapy. However, follow-up has not been performed to see if the reduced risk factors translate into a reduction of the elevated cardiovascular mortality of RA patients.

Ristimäki: What is the target of this antibody treatment? Is it soluble TNF or TNF sitting on the cell?

Feldmann: Both the antibody and the receptors will bind to both membrane TNF and soluble TNF. There is a myth that cells expressing TNF are killed. This is not true *in vivo* although it can be seen *in vitro* at very high (unphysiological) antibody levels using TNF-transfected cells which don't shed TNF.

Brennan: We have recently done studies on the regulation of TNF by the TNF convertase enzyme (TACE). It turns out that membrane TNF is transiently expressed on monocytes. There may well be situations (in disease pathology for example) where the expression of membrane TNF is sustained. This could involve some dysregulation of TACE activity. The genetic regulation of this enzyme has not yet been studied, nor have polymorphic studies been done. It is clear however, that membrane TNF can function like soluble TNF, and this is the case in 'knock-in' mice that express non-cleavable TNF.

Ristimäki: Do these monoclonals get into the synovial fluid and stay there for a long time?

Brennan: Yes, they migrate across the endothelium.

Smyth: Coming back to the epidemiology, in the UK if infliximab is licensed for RA, it will be extremely interesting to know the incidence of cancer in these people over the next decade or two. We are about to start looking at infliximab therapy for TNF-relevant cancers. From an epidemiological perspective, is it possible in the NHS system to have death certification flagged?

Feldmann: This is going to happen spontaneously, because part of the approval process required the keeping of a register of all the patients on biologics. It will certainly happen in the UK. The only problem is that the numbers may be very small. Although NICE have approved this treatment, no extra money goes into the system so that the numbers of patients treated aren't growing very fast as yet.

Balkwill: I wanted to ask about the mechanisms of action of anti-TNF antibodies in Crohn's disease, as opposed to RA. You are getting a clear idea of what is happening in RA, but is it the same in Crohn's?

Brennan: The published data suggest that ulcerative colitis isn't as responsive to anti-TNF treatment as Crohn's disease.

Balkwill: That was another question: why does the anti-TNF antibody (Remicade) work in Crohn's disease but etanercept (soluble TNF receptor fusion protein) does not?

Feldmann: The definitive answers to your questions await a lot of work. My belief as to why Remicade works and etanercept doesn't in Crohn's is related to the mode and dose administration. The antibody is given as an intravenous dose, generating blood levels of about 150–200 µg/ml, which are plateau levels. You get these at day 0, 2 weeks and 6 weeks and then enter a 2 monthly pattern. So patients have had blood levels of 50 µg upwards for probably six weeks. Etanercept is given at a suboptimal dose subcutaneously, so it takes almost 4 weeks to get up to the median of 3 µg/ml. It is just a dose effect, I suspect. In principle there is no reason to believe that intravenous or higher dose etanercept would not work for Crohn's. There are other trials which anecdotally seem to be doing better using other designs and endpoints. In terms of mechanism, the studies performed in Crohn's have not been nearly as extensive as in rheumatoid. While we had a reasonably scientific approach to the anti-TNF studies and their mechanism, with Crohn's the belief is that it is the same mechanism, but the data are not there.

Oppenheim: It is very clear that the use of anti-TNF or the TNF receptor interrupts a cytokine cascade that decreases inflammation in a number of conditions. Kishimoto and colleagues have presented data indicating that anti-IL6 receptor parallels or may even be more effective than anti-TNF (Choy et al 2002). Where does one have to interrupt the network? When you discuss cell contact dependence, and also cells that are conditioned by a cocktail of cytokines, what initiates the process? It seems that perhaps you have studied an ongoing inflammation that is progressing, but if you look at it sequentially earlier in the chain of events, is there any initial cytokine that may play a more major role?

Brennan: The anti-IL6 receptor results are impressive. What seems to be lacking is the mechanism of action. Even Kishimoto doesn't have any good ideas about this, unless it is inducing B cell growth. One possibility is that if you have a very good antagonist — which is where I think the IL6 receptors fall into, as compared to blocking IL1, which has its own problems — the normal homeostatic mechanism may be able to come back in and quell whatever is left. What is clear from our studies is that spontaneous TNF production cannot be modulated in our *ex vivo* systems by blocking all sorts of different cytokines. We have looked at IL15 and IFNγ. What we found to be particularly important for the TNF signal was its contact-dependent signal between T cells and macrophages. We don't know whether this is the event that initiated the inflammation in the beginning. What we are suggesting very clearly is that it is an important step in maintaining the chronicity. The kinetics of the cytokines being produced in these cultures are very different from what occurs in monocytes stimulated with endotoxin.

Oppenheim: Has TNF been over-expressed in any mice? If so, do transgenic over-expressors develop any of these syndromes?

Brennan: George Kollias has done this. We got these mice several years ago and backcrossed them to DBAs. The reason they got an arthritis-like syndrome was unexpected, because deletional mutation of the 3' UTR is important not only for regulation of TNF but also for cell-type expression. Fibroblasts don't normally make TNF protein, but if you mutate the AU-rich sequences or delete the 3' UTR, it facilitates fibroblast expression of TNF protein. In these mice, most of the transgene huTNF is being made in the fibroblasts. Expression of TNF in fibroblasts of the joints will up-regulate a broad range of destructive metalloproteinases resulting in inflammation. It is not the macrophages that are making it, but the fibroblasts. It is an interesting model but it doesn't really relate to what is happening in human disease.

Feldmann: Where Kollias' model is very interesting is that you can backcross these mice to whatever knockout you can find. He has done a lot of this work. TNF transgenics can be backcrossed into RAG knockouts and still get arthritis. Thus once TNF is made, T cells and B cells are no longer important for arthritis. This is not true for inflammatory bowel disease in the same animals. There are differences in different chronic inflammatory diseases, at least in animal models.

Thun: You both pointed out that Remicade is approved in Crohn's disease but not ulcerative colitis. I have a friend with severe ulcerative colitis who has been treated off-label with Remicade and showed a remarkable response. You mentioned that there were ongoing trials with Remicade in ulcerative colitis. Can you tell us why originally it wasn't approved?

Feldmann: One of the commonest reasons for drugs not to be approved is that the wrong trial was done. If you consider how we do animal or *in vitro* experiments, we do a lot until we get the right design. In humans this is not possible. About three trials is the maximum. One ulcerative colitis trial that failed was heroic to the extreme. The end point was prevention of colectomy. The design was to take very severe ulcerative colitis where the standard care would end up being colectomy, and see whether this could be avoided in 50% of these patients. They stopped the trial after 8 patients. When they changed it to going for more common (less severely affected) patients and using an ulcerative colitis disease index, it seemed to be working.

Ristimäki: Mechanistically, in order to increase TNFα expression you need both transcriptional induction and post-transcriptional stabilization of the mRNA. Have you looked at which one of these mechanisms is responsible for NF-κB versus PI3K-mediated TNFα induction in your system?

Brennan: No.

Ristimäki: With respect to COX-2, it has been suggested that NF-κB is the transcriptional inducer and AKT that is downstream of PI3K is the one that is inducing mRNA stability.

Blankenstein: I have a question concerning the bystander cells. Do you want to say they are antigen non-specific?

Brennan: Yes.

Blankenstein: It is difficult to exclude the possibility that there are also specific ones by the methods you use.

Brennan: Yes. The effector response in the rheumatoid T cell population mimics that of our bystander activated T cells. This doesn't mean to say that they are a heterogeneous population of cells, but the predominant effector response is that of a bystander cell.

Blankenstein: Is this a polyclonal or oligoclonal population? What is the percentage of cells that respond to your cytokine cocktail?

Brennan: Years ago, the way to see whether there was a clonal proliferation of a T cell to an unknown antigen was to look for oligoclonal expansion. There isn't a consensus for that occurring in the RA population, although in certain individuals there is evidence for response against collagen type II or gp39. We are suggesting that the large population of the T cells in the joint which have migrated there aren't polyclonally or oligoclonally expanded, but have been maintained in this non-apoptosis state by the cytokine environment in which they find themselves. Stromal cell factors prevent the apoptosis. If the cells were dividing, presumably they would then go on to apoptose, but they aren't — they are just sitting there in an inert but activated state. They aren't making huge amounts of cytokines but they are expressing lots of activation markers. We believe that they are effectors in this contact dependent pathway.

Blankenstein: But the majority of them respond to the cytokines?

Brennan: We don't need to stimulate them. When we take them out of the joint, we don't stimulate them further *ex vivo*. We are using them in their *in vivo* state, and we can mimic them by taking normal blood T cells and activating them with cytokines.

Reference

Choy EH, Isenberg DA, Garrood T et al 2002 Therapeutic benefit of blocking interleukin-6 activity with an anti-interleukin-6 receptor monoclonal antibody in rheumatoid arthritis: a randomized, double-blind, placebo-controlled, dose-escalation trial. Arthritis Rheum 46:3143–3150

How do chemokine/chemokine receptor activations affect tumorigenesis?

Ann Richmond, Guo Huang Fan, Punita Dhawan and Jinming Yang

Department of Veterans Affairs, Nashville, TN and Department of Cancer Biology, Vanderbilt University School of Medicine, Nashville, TN 37232, USA

Abstract. Cells that display chemokine receptors are capable of responding to a gradient of chemokine with a motility response that can translate into a chemotactic response. This continuous response to the chemokine sometimes requires that the chemokine receptor be internalized and recycled back to the membrane. We have shown that ligand activation of the CXC chemokine receptor, CXCR2, results in movement of the receptor into clathrin coated pits, followed by movement into the early endosome, the sorting endosome, then on to the recycling endosome prior to trafficking back into the plasma membrane compartment. Prolonged exposure to saturating concentrations of the ligand results in movement of a large percentage of the receptor into the late endosome and on to the lysosome for degradation. Mutation of the receptor in a manner which impairs receptor internalization by altering the binding of adaptor proteins AP-2 or β arrestin to CXCR2, results in a marked reduction in the chemotactic response. Chemokine receptors also activate multiple intracellular signals that lead to the activation of the transcription factor, nuclear factor κ beta (NF-κB). Transformation is often associated with a constitutive activation of NF-κB, leading to endogenous expression of chemokines and their receptors. This creates an autocrine loop with NF-κB in the activated state, and altered expression of factors that promote tumour angiogenesis and escape from apoptosis. We have shown that the constitutive activation of NF-κB in human melanoma tumours is accompanied by constitutive activation of the NF-κB inducing kinase (NIK) as well as the constitutive activation of AKT. As these factors that modulate the expression of anti-apoptotic factors work together, the tumour cells exhibit enhanced survival and growth. This never ending cycle of activation of NF-κB, leading to enhanced production of chemokines, enhanced activation of AKT and NF-κB, and enhanced transcription of inhibitors of apoptosis and chemokines, is one that has been used to foster the growth of the tumour to the disadvantage of the host. Thus we propose that blocking CXCR2 and/or NF-κB offers potential therapeutic promise for a number of chronic inflammatory conditions and cancers, including malignant melanoma.

2004 Cancer and inflammation. Wiley, Chichester (Novartis Foundation Symposium 256) p 74–91

It is now well understood that chemotactic cytokines, or chemokines as they are now referred to, play a number of biological roles. Chemokines mediate leukocyte trafficking, haematopoiesis, viral infection/parasite infection, angiogenesis, tumorigenesis, wound healing, inflammation, atherosclerosis and chronic inflammatory conditions. These diverse processes are mediated through chemokine ligand binding to the seven transmembrane G protein-coupled chemokine receptors. Tumour growth/development can be positively or negatively affected through the action of chemokine receptors (Richmond 2002, Homey et al 2002). The first mechanism for chemokine negative regulation of tumour growth involves tumour production of certain CC and CXC chemokines which recruit tumour infiltrating lymphocytes (TILs), natural killer cells, dendritic cells and macrophages into the developing tumour to facilitate the immune response to the tumour (Muller et al 2001). For example, as tumour cells proliferate, they often produce MCP-1, MIP-1α and other CC chemokines, which recruit monocytes and potentially immature dendritic cells into the tissue environment of the tumour (Homey et al 2002). These recruited monocytes subsequently produce a number of cytokines such as GM-CSF, TNFα and CD40L which activate and facilitate maturation of the immature dendritic cells (Vicari & Caux 2002). The recruited dendritic cells up-regulate expression of the chemokine receptor CCR7 and take up tumour antigens released by apoptosing tumour cells. Subsequently these fully activated dendritic cells migrate through a chemokine gradient (CCL21) to neighbouring lymph nodes, where they present tumour antigen to T and B lymphocytes. The activated lymphocytes then enter the bloodstream, migrate along a chemokine gradient to the growing tumour, and attempt to destroy the growing tumour (Homey et al 2002, Vicari & Caux 2002). A second way that chemokines retard tumour growth is through angiostatic activity in response to CXC chemokines such as CXCL9, 10 and 11 (mig, IP-10 and ITAC) (Homey et al 2002).

Chemokines play tumorigenic roles by inducing angiogenesis, facilitating the metastasis of tumour cells through chemokine receptor display on tumour cells, allowing these tumour cells to migrate to distant organs which express ligand for those receptors (Strieter 2001). For example, CCR7 and CXCR4 are reported to be expressed on breast cancer cells and tumour cells expressing these receptors migrate to lymph node, lung, or liver, where ligands for these receptors are highly expressed (Muller et al 2001). Tumorigenic/angiogenic chemokines can also act in concert with loss of tumour suppressors to enhance tumour growth. Constitutive expression of mCXCL1 (MIP-2) by melanocytes which exhibit a loss of the tumour suppressor $p16^{INK4a}/p19^{ARF}$ is associated with increased incidence of malignant melanoma in response to DMBA treatment (Yang et al 2001). Finally, chemokine activation of chemokine receptors on tumour cells can activate NF-κB (Yang et al 2001), leading to enhanced expression of chemokines, cytokines, and factors that inhibit apoptosis (IAPs).

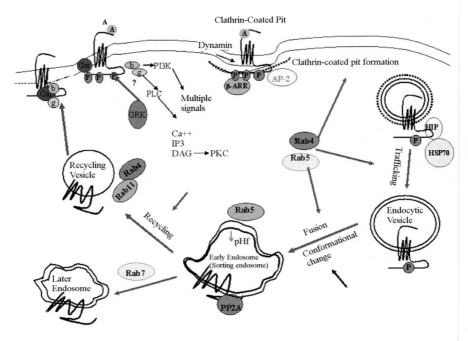

FIG. 1.   Model delineating the pathways activated once ligand binds to CXCR2: this model describes the signalling pathways activated in response to ligand binding to CXCR2, the receptor phosphorylation, internalization, and trafficking through a series of endosomal compartments identified by co-localization of CXCR2 with specific Rab GTPase markers. The receptor internalizes in a Rab5-and dynamin-dependent mechanism associated with the formation of clathrin coated pits as a result of the binding of the adaptor proteins β arrestin and AP-2 to the carboxyl terminal domain of the receptor. As the clathrin lattice forms, the receptor internalizes into a Rab5-containing vesicle. The receptor becomes dephosphorylated by protein phosphatase 2A in the endosomal compartment and this event is thought to be required for the resensitization of the receptor before it recycles back to the membrane in a Rab11-containing endocytic vesicle. Under conditions of exposure to saturating concentrations of ligand as long as 4 h, the internalized receptor moves into the Rab7-containing late endosome and becomes in part targeted for degradation in the lysosome.

With all this continuous migration of leukocytes into the developing tumour in response to chemokines, it is important to consider how the expression of the chemokine receptors is regulated (Fig. 1). For the most part, it is agreed that chemokine binding to chemokine receptors results in the activation of the receptor accompanied by a conformational change in the receptor. With this event, the receptor is available for phosphorylation by G protein receptor kinases (GRKs), and these phosphorylation events are associated with the desensitization of the receptor (Mueller et al 1997). The phosphorylated receptor will subsequently bind β-arrestin molecules which hold the receptor in the desensitized state (Fan et al 2001). Adaptin 2 molecules may also bind to available di-leucine motifs (Fan et al 2001). The binding of these adaptor molecules recruits clathrin to the receptors, and the receptors move into clathrin-coated pits, which pinch off into clathrin-coated vesicles (CCVs) through a dynamin and Rab-dependent mechanism (Yang et al 1999). The CCVs are uncoated before fusing with Rab5-positive early endosomal compartments (Fan et al 2003). Heat shock protein 70 (Hsp70) and its interacting protein, HIP, are postulated to play a role in the uncoating of CCVs (Fan et al 2003). By following the co-localization of CXCR2 with various Rab-containing vesicles in HEK293 cells transiently transfected with fluorescently tagged expression vectors for these proteins, it is possible to see the time dependent transition of CXCR2 through intracellular trafficking vesicles. CXCR2 co-localizes in the clathrin coated vesicles with Rab5 (0.5 to 1 h post ligand stimulation), and the Rab11-associated recycling vesicles (0.5 to 1 h post ligand stimulation). It is while in the early endosome that the receptor is thought to associate with PP2A and undergo dephosphorylation (Fan et al 2001), allowing transport of the receptor to Rab11-positive recycling endosomes, so that when sensitized receptor merges with the Rab11-expressing recycling vesicles, the receptor will be ready to rebind and respond to ligand. This process is called resensitization. If the ligand stimulation remains continuous (>4 h) at saturating concentrations, the receptor moves from the early endosome (Rab5-expressing vesicles) to Rab7-associated late endosomal vesicles. After 4 and 6 h of continuous ligand stimulation it is possible to observe the merging of the Rab7-expressing late endosomes into the lysosomes which are marked by LAMP1 immunostaining. It is here that receptor becomes degraded and permanently down-regulated. By over expressing dominant negative Rab5 in cells expressing CXCR2, ligand induction of receptor internalization is blocked. Moreover, when dominant negative Rab11 is co-expressed with CXCR2, receptor backs up into Rab5- and Rab7-containing endosomes, and does not recycle back to the membrane. Finally, when dominant negative Rab7 is co-expressed with CXCR2, there is an increased shuttling of receptor into Rab5- and Rab11-containing compartments even after prolonged stimulation with high concentrations of ligand (200 ng/ml for 4 h). Altogether, these data demonstrate a pathway for the trafficking of CXCR2 back to the membrane under conditions of non-saturating concentrations of ligand. However, to enable down-modulation of response to chemokines available at high concentrations for continuous periods of time, the receptor is targeted for degradation in the lysosome (Fan et al 2003). In this manner, one might expect that delivery of ligands for CXCR4 or CCR7 to the tumour would block metastasis of tumour cells mediated through these receptors, due to the down-modulation and degradation of the receptors.

Most of the chemokine receptors are internalized subsequent to ligand stimulation and follow a similar pathway for trafficking back to the membrane to allow continued response to a chemokine gradient. However, there is variability in the role of the various adaptor molecules to facilitate this internalization and there are some reports of utilization of caveolae in addition to clathrin-coated pits for movement into the cytoplasmic compartment (Dzenko et al 2001). For CXCR1, CXCR2, CXCR4 it appears that blocking internalization of receptors markedly impairs the ability to respond to a chemotactic gradient mediated by that receptor. On the other hand, mild impairment of the internalization of CCR2b or CCR5 has not been shown to impair chemotactic responses monitored over a 1.5–3 h time period (Arai et al 1997). Certain chemokine receptors (CXCR4 and CCR5) have been demonstrated to undergo ubiquitination followed by degradation by the proteosome after ligand stimulation, and factors that inhibit the proteosome block the chemotactic response to ligands for these receptors (Marchese & Benovic 2001, Fernandis et al 2002). These data support the contention that intracellular trafficking of chemokine receptors may facilitate the cellular response to the extracellular gradient. This response requires conversion and amplification of the pattern of temporal and spatial activation of membrane receptors to establish intracellular polarity and oscillation of this polarization response in a continued fashion. However, the fMLP, C5a and cAMP receptors can apparently mediate chemotaxis in the absence of detectable receptor internalization. It is unclear at this time how this occurs in the absence of receptor internalization. Clearly, it is possible that the main events required to transduce extracellular gradient into appropriate temporal and spatial patterns of intracellular signals are evoked through association of receptor cytoplasmic domains with adaptor proteins, and our modifications of receptor which block chemotactic responses are in fact altering the ability of these adaptors to bind, which not only alters internalization, but also blocks the downstream signals mediated through the binding of these adaptor molecules to the receptor.

Modulation of receptor recycling and degradation remain attractive targets for development of therapeutics for intervening in the inflammatory response. For example, variants of RANTES that fail to dissociate from CCR5 in spite of receptor internalization prevent the continuous response to the ligand, and thus functionally ablate the receptor and its biological effects (He et al 2001).

In the second phase of my discussion I will discuss the role of chemokines as promoters of tumour growth.

- Chemokines can facilitate metastasis by recruiting tumour cells into tissues. This event is based upon the display of chemokine receptors on the tumour cells and organ production of the chemokine ligand for that receptor.

TABLE 1   Chemokine secretion from the cultured melanoma cells

Cell line	CXCL1 (pg/mg)	CXCL8 (pg/mg)	CXCL5 (pg/mg)	IL1β (pg/mg)
NHEM	100 ± 77	63 ± 45	ND	ND
Hs294T	3630 ± 321	34666 ± 1885	40320 ± 3669	191 ± 14
Sk Mel 2	1950 ± 208	1447 ± 27	209 ± 145	ND
Sk Mel 5	19200 ± 1150	340 ± 32	ND	ND
Sk Mel 28	2870 ± 58	3250 ± 288	71 ± 100	ND
WM 115	11500 ± 1030	61250 ± 1767	ND	19 ± 2
WM 164	705 ± 27	492 ± 225	ND	ND
WM 852	77 ± 9	2450 ± 321	35 ± 50	12 ± 11
A 375	2800 ± 595	23197 ± 3689	240 ± 91	38 ± 12

ND, not detectable.
80% confluent Melanoma cells and NHEM cells were incubated in serum-free medium for 24 h at 37 °C and the supernatant was collected and cleared by centrifugation. Aliquots were then subjected to ELISA assay (R&D system) according to the manufacturer's protocol. Cells were lysed in RIPA buffer and assayed for total protein (Bio-Rad). Chemokine levels were calculated as chemokine/cytokine concentration/mg protein in the lysates. (Reprinted with permission from Yang et al 2001.)

- Chemokines can stimulate growth of new blood vessels into the developing tumour.
- Chemokines can regulate transcription of chemokines, metalloproteinases and factors involved in apoptosis.

Chemokines such as CXCL1 and CXCL8 are overexpressed in melanoma and other tumour types (Shattuck et al 1994 and Table 1). The overexpression is related to altered tumorigenic/metastatic processes. How do CXCL1 and CXCL8 facilitate tumorigenesis? One way that these tumorigenic chemokines can facilitate tumour growth is through the stimulation of the growth of blood vessels into the tumour. This activity is antagonized by angiostatic chemokines such as IP-10, mig and ITAC (CXCL9–11). When the receptors for CXCL1 and CXCL8 are expressed on the tumour cells, these ligands can induce a number of intracellular signals which lead to alterations in gene expression. One outcome of these signalling events is activation of the transcription factor NF-κB.

The dysregulation of NF-κB transcription machinery and constitutive expression of chemotactic cytokines are common early events in malignant

TABLE 2   Fold induction of basal luciferase activity from the NF-κB promoter, CXCL1 promoter and AP-1 promoter in cultured melanoma cells

Cell line	NF-κB luciferase	CXCL1 luciferase	AP-1 luciferase
NHEM	1	1	1
Hs294T	5.7±0.1	3.2±0.4	2.5±1.2
Sk Mel 2	5.0±0.8	3.2±1.5	1.5± 0.1
Sk Mel 5	6.2±0.2	5.3±1.0	4.0±0.6
Sk Mel 28	5.5±1.0	1.1±0.5	2.6±0.4
WM 115	6.8±2.8	3.3±1.6	4.0±2.2
WM 164	4.5±0.3	2.2±0.8	4.0±2.1
WM 852	3.4±0.1	2.5±0.5	3.8±0.1
A375	4.8±1.2	5.2±2.5	2.9±0.5

Melanoma cells and NHEM cells with 80% confluence were transfected with the respective constructs. After 48 h of transfection, extracts from cultured cells were prepared, and the activity of luciferase was detected using a Dual-Light kit (Tropix) and normalized by β galactosidase activity. The constitutive induction of promoter activities of NF-κB, CXCL1, and AP-1 by melanoma cells is relative to NHEM. (Reprinted with permission from Yang et al 2001.)

tumour progression (Baldwin 1996). Rel/NF-κB, a family of structurally related DNA-binding proteins, has been implicated in the regulation of cell growth and oncogenesis by inducing proliferative and anti-apoptotic gene products (Beg & Baltimore 1996, Perkins 2000). In non-stimulated cells, NF-κB is sequestered in the cytoplasm and is complexed with IκB, a family of inhibitory proteins, which

---

FIG. 2.   Model of potential components involved in the constitutive activation of NF-κB and enhanced expression of CXCL1 and CXCL8 in melanoma. We postulate that the constitutive activation of NF-κB is modulated by receptor-mediated signals, such as those mediated through chemokine receptors and the LT-β receptor, which result in the persistent activation of NF-κB-inducing kinase (NIK), AKT and potentially, mitogen-activated protein (MAP) kinase kinase kinase 1 (MEKK1). This activation of the inhibitor of NF-κB (I-κB) kinase (IKK; a complex of IKK-α, IKK-β and NEMO) leads to the phosphorylation and degradation of IκB and the phosphorylation of RELA/p65, which can, in turn, lead to constitutive expression of chemokines. We postulate that chemokine expression can be blocked by treatment with non-steroidal anti-inflammatory drugs (NSAIDs), NEMO-binding peptide and PS-341, which would target tumour cells for apoptosis. CBP, CREB-binding protein; CDP, CCAAT displacement protein; HAT, histone acetyltransferase; HDAC1, histone deacetylase 1; IUR, immediate upstream region; Lt-β, lymphotoxin β; LtβR, lymphotoxin β receptor; NEMO, NFκB essential modulator; PARP, poly-ADP ribose polymerase; PI3K, phosphatidylinositol-3-kinase; PtdInsP3, phosphatidylinositol-3,4,5-trisphosphate; PTEN, phosphatase and tensin homologue; TRAF, tumour necrosis factor receptor-associated factor. (Reprinted with permission from Richmond 2002.)

bind to NF-κB and mask its nuclear localization signal, thereby preventing nuclear transport (Baldwin 1996). NF-κB translocation to the nucleus is preceded by IκB phosphorylation (Ser 32 and 36), ubiquitination and ultimately proteolytic degradation. The IκB kinase or IKK complex regulates the IκB phosphorylation and subsequent degradation (Stancovski & Baltimore 1997, Karin 1999, Regnier et al 1997). The IKK complex consists of two catalytic units IKK-α and IKK-β and a regulatory subunit IKK-γ or NEMO (Li et al 2001).

Transient overexpression studies have suggested that some mitogen-activated protein kinase kinase kinases (MAPKKKs), including NF-κB-inducing enzyme

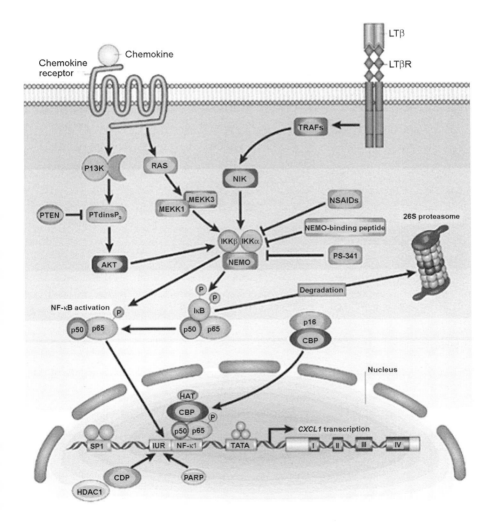

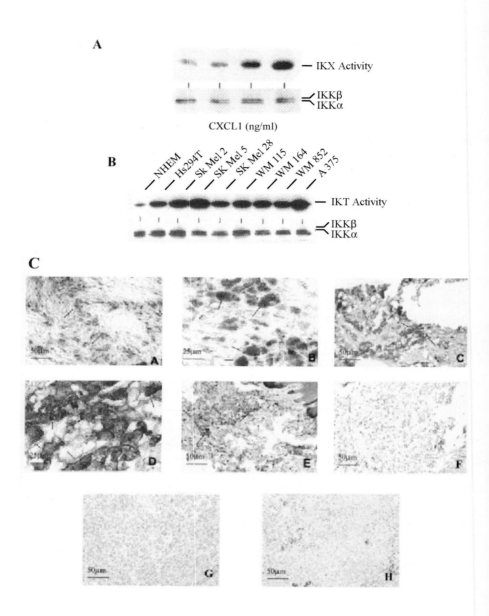

A

— IKX Activity

IKKβ
IKKα

CXCL1 (ng/ml)

B

NHEM Hs294T Sk Mel 2 SK Mel 5 SK Mel 28 WM 115 WM 164 WM 852 A 375

— IKT Activity

IKKβ
IKKα

C

(NIK) and MEKK1-3, are involved in the activation of the IKK complex (Nakano et al 1998, Nemoto et al 1998). NIK was identified by means of its association with TNFR associated factor 2 (TRAF2) and has been shown to activate NF-κB (Malinin et al 1997). The yeast two-hybrid system as well as protein interaction studies have confirmed that NIK interacts with both IKK-α and β, and activates their phosphorylation (Ling et al 1998, Delhase et al 1999). However, recent results from IKK and NIK knockout studies demonstrated that IKK-α and NIK are not required for IKK activation by TNFα, but are required for IKK activation by LT-β (Matsushima et al 2001, Yin et al 2001).

There is a link between the constitutive activation of NF-κB in melanoma cells and the ability of CXCL1 to induce IKK activity in cells expressing CXCR2. Administration of exogenous CXCL1 to cultures of normal human epidermal melanocytes induces the activation of the IKK kinase that phosphorylates the inhibitor of κB, IκB. The activation of this kinase by CXCL1 can be demonstrated to occur in a concentration-dependent manner (Fig. 3A). Moreover, human melanoma tumour cell lines exhibit constitutive activation of IKK. When eight different human melanoma cell lines and cultures of normal human epidermal melanocytes were examined for constitutive IKK activity by immunoprecipitating IKK-α and IKK-β, then testing the ability of the immunoprecipitate to phosphorylate a GST–IκB substrate, we found that the IKK activity is markedly elevated in human melanoma cell lines (three- to 13-fold) as compared to the IKK activity in normal human epidermal melanocytes (Yang & Richmond 2001; Fig. 3B). The constitutive activation of NF-κB in melanoma tumour cells can be further demonstrated by electromobility shift assay where we demonstrate that nuclear extracts of melanoma cells show marked

---

FIG. 3. (A) CXCL1 induces IKK activity in NHEM. The 80% confluent NHEM cells cultured in serum-free 154 medium were stimulated with CXCL1 at the indicated concentrations for 24 h. Protein (400 μg) from cytosolic extracts was immunoprecipitated with IKK-α/β polyclonal antibodies. IKK activity was assayed using recombinant GST-IKBα (amino acids 1–54) as substrate in the presence of [$^{\alpha 32}$P]ATP (*top panel*). (B) The constitutive activation of IκB kinases in melanoma cells was analysed by autoradiography same method as above. Blots of the same membrane were probed with the anti-IKK-α and anti-IKK-β polyclonal antibodies to quantitate and normalize the IKK activities to the total IKK protein immunoprecipitated. (C) Paraffin-embedded sections from biopsy specimen were immunostained with antibody for activated p65. The staining indicates a positive immunolocalization of activated p65. Melanomas stain strongly for activated p65 (example: patient # 19, Panel A and B, 20× and 100× respectively and patient # 25, Panel C and D, 20× and 100× respectively). Severely dysplastic nevi also stained strongly (example: patient # 14, Panel E, 20×). Activated p65 was not observed in nevi with mild dysplasia (example: patient #11, Panel F, 20×). The negative control for immunostaining is shown in G (20×). H shows an example of a melanoma (patient # 23) negative for p65 immunostaining. Arrows indicate the activated p65. (Reprinted with permission from Dhawan et al 2002.)

enhancement of p50/p65 heterodimer and p50/50 homodimers as compared to normal human epidermal melanocytes (Dhawan & Richmond 2002). In addition, transfection of melanoma cells and normal human melanocytes with an NF-κB luciferase reporter construct demonstrates that melanoma cells exhibit enhanced NF-κB activity in comparison to normal melanocytes. When human melanoma lesions, severely dysplastic nevi, dysplastic nevi, and normal nevi are stained with antibody to the NLS of p65, which recognizes that p65 is not bound to IκB, there is increased nuclear staining for p65/RelA in melanoma and severely dysplastic nuclei, as compared to dysplastic and normal nevi (Dhawan & Richmond 2002; Fig. 3C).

This enhanced NF-κB activity is associated with enhanced activity of factors upstream of IKK including NIK and AKT (Dhawan & Richmond 2002, Dhawan et al 2002). Melanoma tumour cells immunoprecipitated with IKK antibody demonstrate enhanced association of NIK as compared to NHEM and vice versa (Dhawan & Richmond 2002). Dominant negative NIK transfected into

ENHANCED IKK ACTIVITY IS ASSOCIATED WITH NIK IMMUNOPRECIPITATED FROM MELANOMA CELLS

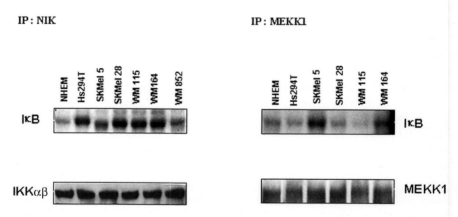

FIG. 4. The level of NIK protein and IKK associated NIK is higher in some melanoma cells. (A) The normal and melanoma cell lines were incubated in serum-free media overnight. The proteins were resolved on a 10% SDS-PAGE under reducing conditions and transferred to nitrocellulose membrane. The membrane was blotted with NIK and MEKK1 antibody and was stripped and reprobed with actin antibody to monitor equal loading of protein. This figure is representative of three separate experiments. (B) Co-immunoprecipitation assays. Whole cell extracts were made from normal and malignant melanoma cells, the MEKK1 was immunoprecipitated and the protein was assayed for IKK activity using the GST-I6B protein as a substrate. The amount of immunoprecipitated MEKK1 in the IP was monitored by Western blot. This figure is representative of three independent experiments.

melanoma cell lines blocks the constitutive NF-κB luciferase reporter activity (Dhawan & Richmond 2002; Fig. 4). In comparison, when the same experiments are performed with MEKK1 antibody, only two of the five melanoma cell lines examined exhibited enhanced IKK activity associated with the MEKK1 immunoprecipitated, compared to control NHEMs. Co-transfection of dominant negative MEKK1 and the NF-κB luciferase reporter construct demonstrated that there was no significant inhibition of the constitutive NF-κB activity in melanoma cells with the dominant negative MEKK1 (Dhawan & Richmond 2002). These data suggest that constitutive NIK expression and association with IKK may be an upstream modulator of the constitutive IKK activity in human melanoma cells. We postulate that the constitutive NIK in melanoma cells could be the result of an autocrine activation through secretion of lymphotoxin β or another cytokine or chemokine. Alternatively, there may be an activating mutation of NIK in the melanoma tumour cells, similar to the activating mutations of B-raf, another MAP-3K, recently noted to be abundant in melanoma (Davies et al 2002). Still a third possibility is that there could be loss of an inhibitory activity for NIK in melanoma, such as the loss of the raf kinase inhibitory protein (RKIP) which negatively regulates NIK (Yeung et al 2001).

The constitutive activation of NIK in tumours is not confined to melanoma. We observed that both breast cancer cell lines and lung cancer cell lines also exhibit enhanced NIK activity as compared to non-transformed breast and lung epithelial cell lines (Dhawan & Richmond 2002). These data suggest that constitutive activation of NIK may prove to be a common event during tumour progression.

While NIK can lead to enhanced IKK activity and increased nuclear transport of the p50/p65 components of the NF-κB transcriptional complex, the p65 subunit also needs to become phosphorylated on serine residues to be maximally active. One kinase reported to be associated with this phosphorylation is AKT. AKT becomes activated in response to the activation of PI3K, which results in the production of phosphatidyl-inositol 3,4,5-phosphate (PIP3). PIP3 recruits PDK1 to the membrane and AKT becomes activated through phosphorylation on serine residues. There are three isoforms of AKT (AKT1, 2 and 3). AKT1 is phosphorylated on S473 when activated, while S472 is phosphorylated when AKT3 is activated. AKT can also be activated by other upstream kinases, including Ca^{2+} and calmodulin-dependent kinase protein kinase kinase. To determine whether AKT activation might be constitutive in melanoma cells and thus contribute to the NF-κB activity in these tumours, we first examined the phosphorylation state of AKT in melanoma cell lines using a phospho-specific antibody to AKT (Dhawan et al 2002). We determined that two of five melanoma cell lines studied exhibited marked constitutive phosphorylation of

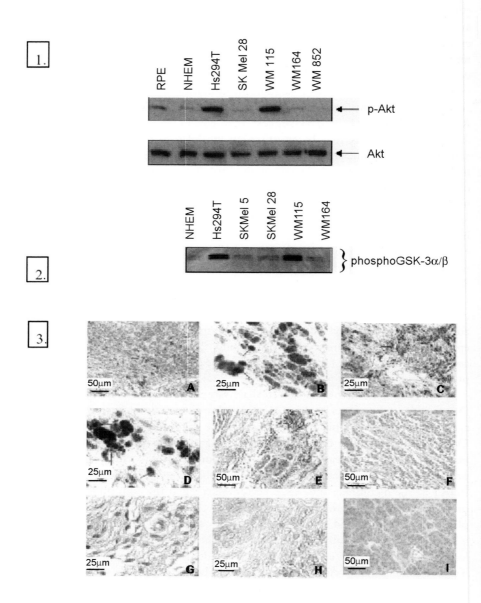

AKT on S473. The enhanced AKT-kinase activity in these samples was confirmed by immunoprecipitation, followed by an AKT kinase activity assay, using the GSK-3 $\alpha\beta$ substrate to monitor AKT activity (Dhawan et al 2002; Fig. 5.2). This enhanced AKT activity correlated with a loss or reduction in the level of the lipid phosphatase, PTEN, known to antagonize the AKT activation (Dhawan et al 2002).

While only two of the five melanoma cell lines studied exhibited enhanced AKT activity, when a group of pigmented lesions was surveyed for phospho-AKT immunostaining, we observed that eight of 11 human melanoma lesions strongly stained for phospho-AKT, while normal nevi and dysplastic nevi exhibited little staining for phospho-AKT (Dhawan et al 2002; Fig. 5.3). In contrast, the severely dysplastic nevi stained strongly for phospho-AKT. Diagnostic markers for these lesions are difficult to find, and our data suggest that characterization of the levels of phospho-AKT in dysplastic nevi might be predictive of the level of progression toward malignancy for the lesion. We are developing long-term studies to examine the utility of phospho-AKT as a marker for malignant progression in melanoma. The activation of AKT is associated with blockage of caspase activity as well as activation of NF-κB which leads to enhanced transcription of inhibitors of apoptosis. Thus AKT activation can promote the survival of tumour cells. By developing mechanisms for blocking this activity, better chemotherapeutic intervention should be possible.

FIG. 5.   AKT is constitutively activated in Hs294T and WM115 melanoma cell lines. Normal (NHEM) and melanoma cells were cultured in absence of serum overnight and the next morning whole cell extracts were prepared for immunoblotting. (1) The expression of both activated and total AKT was determined using anti-phospho AKT (Ser473) and anti-AKT antibodies. (2) *In vitro* kinase assays of AKT immunoprecipitates from normal and melanoma cells. AKT activity was determined using AKT kinase assay with GSK-3α/β as a substrate. This figure is representative of three separate experiments. (3). Paraffin-embedded sections from biopsy specimen of patients were immunostained with antibody for activated AKT. The staining indicates a positive immunolocalization of AKT phosphorylated on S473. Melanomas stain strongly for activated AKT (example: patient # 19, Panel A and B, 20× and 100×, respectively, and patient # 25, Panel C and D, 20× and 100× respectively). Severely dysplastic nevi also stain strongly (example: patient # 14, E, 20×). Activated AKT was not observed in nevi with mild dysplasia (example: patient # 11, Panel F is 20× and G is 100×). The negative control for immunostaining in B (patient # 19) is shown in H (100×). Panel I shows an example of a melanoma with no staining for AKT (patient # 23, 20×). The brown colour (seen here in grey) is due to the pigment in the tumour cells and does not represent activated AKT. Arrows indicate the p-AKT staining. 800 mg of protein from each sample was used for immuno-precipitation using IKK-α and -β antibody (2 mg) to monitor the endogenous levels. These immunoprecipitated extracts were then used to monitor the NIK or MEKK1 associated with IKK-$\alpha\beta$ by Western blot. The same membrane was blotted with IKK-$\alpha\beta$ for normalization. This figure is a representative of three different experiments. (Reprinted with permission from Dhawan et al 2002.)

References

Arai H, Monteclaro FS, Tsou C-L, Franci C, Charo IF 1997 Dissociation of chemotaxis from agonist induced receptor internalization in a lymphocyte cell line transfected with CCR2B. Evidence that directed migration does not require rapid modulation of signaling at the receptor level. J Biol Chem 272:25037–25042

Baldwin AS Jr 1996 The NF-kappa B and I kappa B proteins: new discoveries and insights. Annu Rev Immunol 14:649–683

Beg AA, Baltimore D 1996 An essential role for NF-kappa B in preventing TNF-alpha-induced cell death. Science 274:782–784

Davies H, Bignell GR, Cox C et al 2002 Mutations of the BRAF gene in human cancer. Nature 417:949–954

Delhase M, Hayakawa M, Chen Y, Karin M 1999 Positive and negative regulation of IkappaB kinase activity through IKKbeta subunit phosphorylation. Science 284:309–313

Dhawan P, Richmond A 2002 A novel NFκB-inducing-kinase-MAPK signaling pathway upregulates NFκB activity in melanoma cells. J Biol Chem 277:7920–7928

Dhawan P, Singh A, Ellis D, Richmond A 2002 Constitutive activation of AKT/protein kinase B in melanoma leads to up-regulation of NFκB and tumor progression. Cancer Res 62:7335–7342

Dzenko KA, Andjelkovic AV, Kuziel WA, Pachter JS 2001 The chemokine receptor CCR2 mediates the binding and internalization of monocyte chemoattractant protein-1 along brain microvessels. J Neurosci 21:9214–9223

Fan GH, Yang W, Wang D, Qian Q, Richmond A 2001 Identification of a motif in the carboxylterminus of CXCR2 involved in adaptin-2 binding and receptor internalization. Biochemistry 40:791–800

Fan GH, Lapiere L, Goldenring J, Richmond A 2003 Differential regulation of CXCR2 trafficking by Rab GTPases. Blood 101:2115–2124

Fernandis AZ, Cherla RP, Chernock RD, Ganju RK 2002 CXCR4/CCR5 down-modulation and chemotaxis are regulated by the proteasome pathway. J Biol Chem 277: 18111–18117

He R, Browning DD, Ye RD 2001 Differential roles of the NPXXY motif in formyl peptide receptor signaling. J Immunol 166:4099–4105

Homey B, Muller A, Zlotnik A 2002 Chemokines: agents for the immunotherapy of cancer? Nat Rev Immunol 2:175–184

Karin M 1999 How NF-kappa B is activated: the role of the I kappa B kinase (IKK) complex. Oncogene 18:6867–6874

Li XH, Fang X, Gaynor RB 2001 Role of IKK gamma/nemo in assembly of the I kappa B kinase complex. J Biol Chem 276:4494–4500

Ling L, Cao Z, Goeddel DV 1998 NF-kappaB-inducing kinase activates IKK-alpha by phosphorylation of Ser-176. Proc Natl Acad Sci USA 95:3792–3797

Malinin NL, Boldin MP, Kovalenko AV, Wallach D 1997 MAP3K-related kinase involved in NF-kappa B induction by TNF, CD95 and IL-1. Nature 385:540–544

Marchese A, Benovic JL 2001 Agonist-promoted ubiquitination of the G protein-coupled receptor CXCR4 mediates lysosomal sorting. J Biol Chem 276:45509–45512

Matsushima A, Kaisho T, Rennert PD et al 2001 Essential role of nuclear factor (NF)-kappaB inducing kinase and inhibitor of kappaB (IkappaB) kinase alpha in NF-kappa B activation through lymphotoxin beta receptor, but not through tumor necrosis factor receptor I. J Exp Med 193:631–636

Mueller SG, White JR, Schraw WP, Lam V, Richmond A 1997 Ligand induced desensitization of the human CXC chemokine receptor-2 modulated by multiple serine residues in the carboxyl terminal domain of the receptor. J Biol Chem 272:8207–8214

Muller A, Homey B, Soto H et al 2001 Involvement of chemokine receptors in breast cancer metastasis. Nature 410:50–56

Nakano H, Shindo M, Sakon S et al 1998 Differential regulation of I kappa B kinase alpha and beta by two upstream kinases, NF-kappaB-inducing kinase and mitogen-activated protein kinase/ERK kinase kinase-1. Proc Natl Acad Sci USA 95:3537–3542

Nemoto S, DiDonato JA, Lin A 1998 Coordinate regulation of I kappa B kinases by mitogenactivated protein kinase kinase kinase 1 and NF-kappaB-inducing kinase. Mol Cell Biol 18:7336–7343

Perkins ND 2000 The Rel/NF-kappa B family: friend and foe. Trends Biochem Sci 25:434–440

Regnier CH, Song HY, Goeddel DV, Cao Z, Rothe M 1997 Identification and characterization of an I kappa B kinase. Cell 90:373–383

Richmond A 2002 NFκB, chemokine gene transcription and tumour growth. Nat Rev Immunol 2:664–674

Shattuck RL, Wood LD, Jaffe GJ, Richmond A 1994 MGSA/GRO transcription is differentially regulated in normal retinal pigment epithelial and melanoma cells. Mol Cell Biol 14:791–802

Stancovski I, Baltimore D 1997 NF-kappa B activation: the I kappa B kinase revealed? Cell 91:299–302

Strieter RM 2001 Chemokines: not just leukocyte chemoattractants in the promotion of cancer. Nat Immunol 2:285–286

Vicari AP, Caux C 2002 Chemokines in cancer. Cytokine Growth Factor Rev 13:143–154

Yang JM, Richmond A 2001 Constitutive IKK activity correlates with NFκB activation in human melanoma cells. Cancer Res 61:4901–4909

Yang W, Wang D, Richmond A 1999 Role of clathrin-mediated endocytosis in CXCR2 sequestration, resensitization and transduction. J Biol Chem 274:11328–11333

Yang JM, Luan J, Yu Y et al 2001 Induction of melanoma in murine macrophage inflammatory protein 2 transgenic mice heterozygous for inhibitor of kinase/alternate reading frame. Cancer Res 61:8150–8157

Yin L, Wu L, Wesche H et al 2001 Defective lymphotoxin-beta receptor-induced NF-kappaB transcriptional activity in NIK-deficient mice. Science 291:2162–2165

Yeung KC, Rose DW, Dhillon AS et al 2001 Raf kinase inhibitor protein interacts with NF-kappaB inducing kinase and TAK1 and inhibits NF-kappaB activation. Mol Cell Biol 21:7207–7217

DISCUSSION

Feldmann: What is hooked up to NIK? Currently, it is believed to be lymphotoxin, CD40L and BAFF receptors. Is there anything else?

Richmond: Some data suggest that CCR7 can activate NIK. The CCR7 knockouts show abnormalities in NIK activation. But lymphotoxin β is clearly one of the potential players. Many melanoma cells make this and express the receptors for it. It is certainly possible that there could be an autocrine loop through lymphotoxin β. Also there is potential for an autocrine loop to activate NIK through the chemokine receptors. CCR7 is expressed on a number of melanoma tumours as well. This could be a potential avenue for the activation of NIK.

Ristimäki: You have nicely shown that there is a distinction between dysplastic nevi and malignant melanoma. There are two growth patterns in melanoma: the radial and the vertical. Can you see any distinction between these two types?

Richmond: We have a limited number of primary melanoma lesions. We do have severely dysplastic nevi, which may be interpreted by some as primary melanomas or melanomas *in situ*. I will need to get back with the pathologist to determine whether these are vertical growth phase or radial growth phase. All the other melanomas we examined were metastatic.

Balkwill: You have shown very nice staining for AKT in the sections. If you look in the sections for the chemokine receptors or the chemokines that you are studying, do you find them on all the cells or is it a subpopulation?

Richmond: Subpopulations. It varies from tumour to tumour. A couple of years back we published work looking at IP10, CXCR2 and chemokine staining (Luan et al 1997). There were some lesions with high levels of CXCR2 receptor expression over a number of the tumour cells, while other lesions showed CXCR2 staining consistently in the endothelial cells, and to a lesser extent over subpopulations of the tumour cells.

Balkwill: Do you tend to find the chemokine itself in the same cells that express the receptor, *in situ* in the real tumour?

Richmond: It is hard to say, because of the secretion problem. But there is a low level of expression of the chemokine in most of the tumour cells, and higher levels in certain cells. I would say in general that there is both ligand and receptor expression.

Balkwill: That is interesting. It is analogous to what we see in ovarian cancer, where the receptor is on the minority of cells but the chemokine ligand is expressed by the majority.

Pepper: Does the repertoire of chemokines and their receptors change between primary and metastatic tumours?

Richmond: It would be nice to know this. Perhaps Al Zlotnik's work might answer this better than ours. He has used RT-PCR to look at the repertoire of chemokine receptors expressed on different types of lesions, but I haven't seen this published yet.

Feldmann: What does blocking NF-κB do to the melanoma cells? Could this be a link between inflammation and cancer?

Richmond: We have used inhibitors of NF-κB to block tumour growth. We have used PS341, a proteosome inhibitor, to block NF-κB and other events. When we use this in combination with chemotherapeutic agents, we see beautiful regression of tumours.

Brennan: Do you see apoptosis?

Richmond: Yes, we see marked apoptosis.

Mantovani: I have a very general question. You have studied melanoma thoroughly. Do you think that CXCR2 is a valuable therapeutic target in melanoma?

Richmond: Yes I do think that CXCR2 will be a good therapeutic target for inhibition of tumour angiogenesis and tumour growth. It is my understanding that the studies performed with humanized anti-IL8 antibodies worked fairly well to inhibit melanoma tumour growth in the mouse model, but not so well in a clinical trial. It would be much more effective to use an antagonist to the receptor rather than the ligand. There are five or six ligands that are going to be able to bind to those receptors. In chemokine profiles on melanomas we see that CXC ligand 1 and CXC ligand 8 are both being produced, so it is better to go after the receptor.

Strieter: I would advocate using an antagonist to angiogenesis rather than trying to antagonize the ligand or the receptor of a proangiogenic molecule.

Reference

Luan J, Shattuck-Brandt R, Haghnegahdar H et al 1997 Mechanism and biological significance of constitutive expression of MGSA/GRO chemokines in malignant melanoma tumor progression. J Leukoc Biol 62:588–597

Proinflammatory cytokines, immune response and tumour progression

Michela Spadaro and Guido Forni[1]

Department of Clinical and Biological Sciences, University of Torino, Ospedale San Luigi Gonzaga, 10043 Orbassano, Italy

Abstract. Tumour cells naturally secrete proinflammatory cytokines and chemokines to interact with the microenvironment and regulate neoangiogenesis. The repertoire of factors thus produced shapes tumour progression. However, experiments in the mouse have shown that injection of a low pharmacological dose of a proinflammatory cytokine or chemokine into the microenvironment increases the inflammatory reaction so enormously that locally activated leukocytes inhibit or eradicate the tumour. Massive shrinkage of recurrent head and neck squamous cell carcinomas (SCC) and prevention of recurrences after surgical removal of a primary SCC follow perilymphatic administration of low doses of interleukin (IL)2, while low daily doses of IL12 markedly delay carcinogenesis in transgenic mice predestined to die of mammary carcinomas and keep them tumour-free for long periods. The reaction elicited by proinflammatory cytokines evidently has a great potential in tumour control.

2004 Cancer and inflammation. Wiley, Chichester (Novartis Foundation Symposium 256) p 92–105

The influence of an inflammatory response on tumour growth was recognized more than 120 years ago by Virchow (Balkwill & Mantovani 2001), though whether it will be enhanced or inhibited is a highly controversial subject. Neutrophils, for example, are typical inflammatory cells. Their local presence, however, may be associated with either the growth (Pekarek et al 1995) or the efficient rejection (Di Carlo et al 2001) of a tumour. A similarly clear-cut contrast has been observed for both macrophages (Mantovani et al 1992) and lymphocytes infiltrating the growth area (Mihm et al 1996). The literature on the influence of inflammation on tumour progression, or on the effect of cytokines and chemokines on tumour growth and regression provides a similar dichotomy. The clinical data, too, show that inflammation promotes tumour growth and its inhibition reduces

---

[1]This paper was presented at the symposium by Guido Forni to whom all correspondence should be addressed.

the risk of cancer, but also that a powerful antitumour response rests on the inflammatory reaction activated by cytokines and other proinflammatory agents.

The source of this contrast, however, is purely semantic. The same term, in fact, is applied to two distinct sets of reactions on the part of the same cells performing in different scenarios (Musiani et al 1997) marked by enormous differences in the amount of the cytokine or chemokine (here referred to as cytokines) promoting the inflammatory reaction, the way it is released, and the intensity of the inflammation. The tumour-enhancing effect of a mild inflammatory reaction sluggishly activated by an incipient tumour is only apparently similar to the abrupt start of a violent reaction activated by a single cytokine arriving in amounts various log levels higher than the cocktail of cytokines spontaneously released by tumour cells.

The inflammatory reaction often naturally associated with the growth of a tumour mostly rests on the ability of its cells to release various cytokine cocktails. After lymphoid cells, tumour cells are the most active producers of cytokines and tumours themselves can be characterized by the function of the repertoire of factors they produce (Pekarek et al 1993). These cytokines are instrumental for the interactions between the tumour and microenvironment, the recruitment of infiltrating inflammatory cells and their activation to produce downstream factors and contribute to the switch of tumour angiogenesis.

Cytokine and chemokine-activated tumour inhibition

The mild inflammatory reaction naturally elicited by a tumour can be radically modified by 'pharmacological' amounts of a proinflammatory cytokine (Forni et al 1985). Cytokines repeatedly injected at the growth site of a tumour or around its draining lymph nodes, and cytokines released by cytokine-gene engineered tumour cells, recruit and activate inflammatory cells to elicit a reaction that is often powerful enough to eradicate an incipient tumour and induce the establishment of an efficient tumour-specific immune memory (Musiani et al 1997). Unexpectedly, the mechanisms of cytokine-activated inflammation have proved different from those that might have been predicted from the *in vitro* data (Forni et al 1987). Tumour rejection does not rest only on the ability of local cytokines to restore the activity of lymphocytes trapped in the immunosuppressive tumour microenvironment, but stems from a complex inflammatory reaction leading to both eradication and induction of an efficient, long-lasting, systemic, T cell-dependent immune memory (Colombo et al 1992).

Four general conclusions can be drawn from the data on inflammatory reactions activated by the local injection of pharmacological levels of a great variety of cytokines:

(1)   The quick and non-specific inflammatory reaction elicited inhibits the growth of a tumour and often leads to its eradication.
(2)   Inhibition is the result of direct killing of tumour cells by effector leukocytes recruited and maximally activated by the cytokine released, or by the downstream cytokines it induces.
(3)   The leukocytes thus recruited vary in the function of their cytokines.
(4)   Cytokines repeatedly injected at the growth site of a tumour and those released by cytokine-gene engineered tumour cells elicit a similar reaction. Engineered tumour cells behave like biological cytokine pumps.

The progression of cytokine-activated inflammation has been monitored following a challenge with tumour cells engineered to release proinflammatory cytokines. During the first 8–72 h, the engineered cells behave like their wild-type parental cells and give rise to a solid tumour aggregate whose growth is accompanied by local accumulation of the secreted cytokine to a pharmacologically active concentration. Since most tumours do not express receptors for such cytokines, autocrine effects are not prominent. Instead, after 1–5 days many locally recruited leukocytes penetrate the tumour and kill its cells. T cells are too few to cause any significant destruction on their own at this stage. Their cross-talk with the inflammatory cells suggests that they act as guides through the intense release of secondary cytokines, as revealed by *in situ* hybridization experiments (Colombo et al 1992). The importance of this cross-talk is illustrated by the observation that the antitumour reaction is marginal in mice selectively deprived of either T cells or polymorphonuclear granulocytes (Musiani et al 1997). Progressive killing of the engineered cells and replacement of the tumour itself by granulation tissue full of macrophages or other antigen presenting cells (Forni et al 1993) that have ingested the cell debris constitute the final stage of the reaction (days 5–10). Draining nodes grow larger and display expanded cortical and paracortical areas, as well as numerous tingible-body macrophages and granulocytes. This loading of the immune system with tumour debris, combined with the inflammatory setting built up by the secreted cytokine, creates an appropriate environment for the indirect presentation of tumour antigens to $CD4^+$ and $CD8^+$ memory T cells.

Establishment of unexpected memory responses against poorly or non-immunogenic tumours seems to be an outcome of the unique situation created during cytokine-activated tumour rejection (Allione et al 1994). This enhanced immunogenicity is probably not directly attributable to the cytokine released, but the outcome of four factors directly or indirectly associated with its biological activity:

(1)   The way in which the engineered tumour cells die.

(2)   Loading of the immune system with a large amount of tumour antigens.
(3)   Recruitment and activation of the appropriate antigen presenting cells.
(4)   The cytokine's own immunological role.

If the cytokine leads to quick tumour cell destruction, there is not enough tumour antigen to trigger an effective indirect presentation by host antigen presenting cells and no memory is elicited. An interesting feature is that the tumour begins by expanding. Afterwards, the kinetics of its rejection depends on the efficacy of the inflammatory reaction it provokes. Slow reactions allow a tumour to continue to grow before it is rejected. Cytokines that elicit a sound immune memory against a non-immunogenic tumour cause the appropriate regression of engineered cells that have achieved a certain degree of growth. This explains why the protection afforded by vaccination with engineered tumour cells whose proliferative potential has been inhibited is often not significantly higher than that elicited by wild-type parental cells (Allione et al 1994).

Of even greater interest is the observation that cytokines released by engineered tumour cells control the promotion of particular memory mechanisms. A vigorous T cell-mediated cytotoxicity is generated when rejection is elicited by release of the Th1 cytokines IL2, IL7 and interferon (IFN)γ, whereas the development of cytotoxic T lymphocytes is inhibited and the memory is mediated by non-cytotoxic $CD8^+$ lymphocytes and antibodies in the presence of the Th2 cytokine IL4 (Musiani et al 1997).

Deflection of memory mechanisms towards a Th1 or Th2 response may result from direct influence of the cytokine on memory T cells. It can also depend on its ability to recruit distinct leukocyte populations, prompt presentation by different sets of antigen presenting cells and induce the release of characteristic repertoires of secondary cytokines. Inhibition of a subsequent challenge is conferred by both Th1 and Th2 reactivity. Regression of already established tumours, on the other hand, is due to Th1 reactivity only.

Local IL2 induces the regression of head and neck squamous cell carcinomas (SCCs) and extends survival

The unique ability of the cytokine-activated reaction to eradicate incipient mouse tumours indicated that a similar approach might result in regression of human tumours. For ethical reasons, poor prognosis patients with recurrent SCCs were first treated with low pharmacological doses of IL2 injected around the cervical tumour draining lymph nodes (Cortesina et al 1988, Musiani et al 1989, Cortesina et al 1991) . Several complete or partial responses were documented clinically, radiologically and pathologically in patients with small but inoperable tumours. Surprisingly shrinkage was also induced by low doses of recombinant IL2 injected 1.5 cm from the insertion of the sternocleidomastoid muscle on the mastoid, but not in patients treated with higher doses. Unfortunately, these tumours always recurred after 3–5 months and were poorly responsive to further IL2 treatments (Cortesina et al 1994).

Those results were achieved in patients judged unlikely to respond to long periods of conventional management. Their immune reactivity was marginal because the marked immunosuppression induced by an SCC barred the establishment of long-lasting systemic immunity. Smaller tumours prior to surgery and minimal residual disease may be more promising targets. The prognosis of patients with SCCs is poor since local recurrences or metastases in the cervical lymph nodes usually appear within two years. A randomized, multicentre Phase III trial was therefore conducted to determine whether the disease-free interval and overall survival of patients with T2-T4,N0-N3,M0 SCC of the oral cavity or oropharynx could be extended by combining surgery (and radiotherapy, if required) with perilymphatic IL2. Five thousand units of IL2 were injected around the ipsilateral cervical lymph node chain daily for 10 days before surgery. After surgery, contralateral 5-day IL2 courses were administered monthly for 1 year. The results in the 202 patients who completed the study showed that IL2 did not give rise to significant complications, and that surgery and radiotherapy were not hampered by its prior administration. Multivariate analysis conducted to determine the extent to which survival was influenced by IL2 and the other variables showed that IL2 significantly lengthened disease-free survival and that this resulted in longer overall survival (De Stefani et al 2002).

A fair conclusion is that the reaction activated by a low pharmacological dose of IL2 administered perilymphatically significantly delays the recurrence of SCC.

Cytokines in tumour prevention

The clinical and experimental data obtained when proinflammatory cytokines were administered at the site of a tumour or around its draining lymph nodes indicated that the nonspecific reaction thus elicited results in major tumour shrinkage and inhibition of recurrences. Comparative studies showed that the reaction elicited by local IL12 was particularly effective in tumour inhibition (Cavallo et al 1997). Surprisingly, a similar IL12-triggered reaction also inhibited carcinogenesis in mice injected subcutaneously with 3-methylcholanthrene (Noguchi et al 1998).

We therefore decided to determine whether IL12 can also inhibit spontaneous carcinogenesis in mice transgenic for the rat HER-2/neu oncogene and genetically predestined to develop lethal mammary carcinomas. In effect, with a slightly asynchronous, but consistent progression all ten mammary glands of BALB/c mice transgenic for the transforming activated oncogene (BALB-neuT), and one or two glands of FVB females carrying the protoncogene (FVB-neuN) progress through atypical hyperplasia to invasive lobular carcinoma. This progression begins in BALB-neuT mice when they are weaning, but in FVB-neuN mice when they are adults. Systemic treatment of initial preneoplastic lesions with IL12, 5 days a week (1 week on, 3 weeks off; first course 50 ng IL12/day, the remainder 100 ng/day) markedly delayed tumour onset and reduced tumour multiplicity (Boggio et al 1989). A similar inhibition was achieved when 7 week old BALB-neuT mice with fully blown atypical hyperplasia received sixteen weekly injections of 100 ng IL12. This lower-dose and later administration induced high and sustained levels of serum IFNγ equivalent to those elicited by more frequent administrations (Cifaldi et al 2001). In both BALB-neuT and FVB-neuN mice, inhibition was associated with mammary infiltration by reactive cells, production of cytokines and iNOS, reduction in microvessel number and a high degree of haemorrhagic necrosis. These results suggest that IL12 effectively inhibits mammary carcinogenesis when its administration accompanies the passage from atypical hyperplasia to invasive carcinoma. The poor efficacy of late treatments depends on the lower sensitivity of differentiated blood vessels. An assessment was also made of the mechanisms of this anti-angiogenic activity. Spleen cells activated in the presence of IL12 release factors that arrest the cycle of endothelial cells, inhibit *in vitro* angiogenesis, negatively modulate the production of matrix metalloproteinase 9 and the ability of endothelial cells to adhere to vitronectin, and up-regulate ICAM1 and VCAM1 expression. These effects do not require direct cell–cell contact, but result from continuous interaction between activated lymphoid cells and endothelial cells. IP-10 and MIG are pivotal in inducing these effects (Strasly et al 2001). CD4$^+$, CD8$^+$ and NK cells are all needed to mediate the anti-angiogenetic effect of IL12. Moreover, T-lymphocytes from normal and IFNγ knockout mice activated by anti-CD3 and anti-CD28 antibodies and cultured in inserts in the presence of IL12 modulate the genetic programs of tumour lines growing at the bottom of transwells. cDNA gene expression array and RT-PCR, and protein expression showed that LPS-R, TTF1, TGF and FGF genes were up-modulated by factors other than IFNγ released by activated lymphocytes. The high levels of IFNγ released by normal IL-12 activated lymphocytes up-modulated the expression of STAT-1, IRF-1, LMP-2, LMP-7, MIG, MCP-1 and Ang-2 genes, but down-modulated that of VEGF. PA-28, IP-10, iNOS and MIP-2 genes were up-modulated by factors released only by IL12-activated lymphocytes apart from IFNγ. The opposite modulations of VEGF expression and of Ang-2, MIG, IP-10, and iNOS by IL12 activated lymphocytes fit in well with the inhibition of angiogenesis that characterizes IL12's antitumour activity. By directly influencing tumour cells through the cytokines they release, lymphoid cells activated in the presence of IL12 also modulate the growth of a tumour by changing its genetic programs (Cavallo et al 2001). This new way whereby non-specific reactions activated by cytokines affect the growth of a tumour so that it becomes a party to its own inhibition is an unexpected and indeed provocative finding (Cavallo et al 2001).

Conclusions

The advantage of non-specific antitumour responses elicited by proinflammatory cytokines is that they can be directly applied to a broad range of individuals, irrespective of the type of tumour associated antigen their foreseeable tumours may express. The reaction elicited leads to unexpected shrinkage of large tumour masses, certainly larger than those that can be managed by the optimally activated specific immunity. Selective delivery of cytokines at the tumour site is the major issue.

Moreover, the ability of nonspecific cytokine-activated reactions to prevent tumours points to new ways of managing preneoplastic lesions. Selection of not-yet patients and healthy individuals eligible for cytokine-based immunoprevention depends on the toxicity of the treatment envisaged, its length and effectiveness. A cytokine stimulation could be considered for not-yet patients with a genetic risk of cancer, individuals exposed to high carcinogen dose, patients with a preneoplastic lesion and those that probably have minimal residual disease after conventional treatment. Persons at risk for cancer or with a preneoplastic lesion may form a category for which nonspecific immunoprevention could be considered as an alternative to current controversial and distasteful preventive manoeuvres (Cavallo et al 2001).

References

Allione A, Consalvo M, Nanni P et al 1994 Immunizing and curative potential of replicating and nonreplicating murine mammary adenocarcinoma cells engineered with interleukin (IL)-2, IL-4, IL-6, IL-7, IL-10, tumour necrosis factor alpha, granulocyte-macrophage colony-stimulating factor, and gamma-interferon gene or admixed with conventional adjuvants. Cancer Res 54:6022–6026

Balkwill F, Mantovani A 2001 Inflammation and cancer: back to Virchow? Lancet 357:539–545

Boggio K, Nicoletti G, Di Carlo E et al 1998 Interleukin 12-mediated prevention of spontaneous mammary adenocarcinomas in two lines of Her-2/neu transgenic mice. J Exp Med 188: 589–596

Cavallo F, Signorelli P, Giovarelli M et al 1997 Antitumour efficacy of adenocarcinoma cells engineered to produce interleukin 12 (IL-12) or other cytokines compared with exogenous IL-12. J Natl Cancer Inst 89:1049–1058

Cavallo F, Di Carlo E, Quaglino E et al 2001 Prevention by delay: nonspecific immunity elicited by IL-12 hinders Her-2/neu mammary carcinogenesis in transgenic mice. J Biol Regul Homeost Agents 15:351–358

Cifaldi L, Quaglino E, Di Carlo E et al 2001 A light, nontoxic interleukin 12 protocol inhibits HER-2/neu mammary carcinogenesis in BALB/c transgenic mice with established hyperplasia. Cancer Res 61:2809–2812

Colombo MP, Modesti A, Parmiani G, Forni G 1992 Local cytokine availability elicits tumour rejection and systemic immunity through granulocyte-T-lymphocyte cross-talk. Cancer Res 52:4853–4857

Cortesina G, De Stefani A, Giovarelli M et al 1988 Treatment of recurrent squamous cell carcinoma of the head and neck with low doses of interleukin-2 injected perilymphatically. Cancer 62:2482–2485

Cortesina G, De Stefani A, Galeazzi E et al 1991 Interleukin-2 injected around tumour-draining lymph nodes in head and neck cancer. Head Neck 13:125–131

Cortesina G, De Stefani A, Galeazzi E et al 1994 Temporary regression of recurrent squamous cell carcinoma of the head and neck is achieved with a low but not with a high dose of recombinant interleukin 2 injected perilymphatically. Br J Cancer 69:572–576

De Stefani A, Forni G, Ragona R et al 2002 Improved survival with perilymphatic interleukin 2 in patients with resectable squamous cell carcinoma of the oral cavity and oropharynx. Cancer 95:90–97

Di Carlo E, Forni G, Lollini P, Colombo MP, Modesti A, Musiani P 2001 The intriguing role of polymorphonuclear neutrophils in antitumour reactions. Blood 97:339–345

Forni G, Giovarelli M, Santoni A 1985 Lymphokine-activated tumour inhibition in vivo. I. The local administration of interleukin 2 triggers nonreactive lymphocytes from tumour-bearing mice to inhibit tumour growth. J Immunol 134(2):1305–1311

Forni G, Giovarelli M, Santoni A, Modesti A, Forni M 1987 Interleukin 2 activated tumour inhibition in vivo depends on the systemic involvement of host immunoreactivity. J Immunol 138:4033–4041

Forni G, Giovarelli M, Cavallo F et al 1993 Cytokine induced tumour immunogenicity: from exogenous cytokines to gene therapy. J Immunother 14:253–257

Mantovani A, Bottazzi B, Colotta F, Sozzani S, Ruco L 1992 The origin and function of tumour-associated macrophages. Immunol Today 13:265–270

Mihm MC Jr, Clemente CG, Cascinelli N 1996 Tumour infiltrating lymphocytes in lymph node melanoma metastases: a histopathologic prognostic indicator and an expression of local immune response. Lab Invest 74:43–47

Musiani P, De Campora E, Valitutti S et al 1989 Effect of low doses of interleukin-2 injected perilymphatically and peritumourally in patients with advanced primary head and neck squamous cell carcinoma. J Biol Response Mod 8:571–578

Musiani P, Modesti A, Giovarelli M et al 1997 Cytokines, tumour-cell death and immunogenicity: a question of choice. Immunol Today 18:32–36

Noguchi Y, Jungbluth A, Richards EC, Old LJ 1998 Effect of interleukin 12 on tumour induction by 3-methylcholanthrene. Proc Natl Acad Sci USA 93:11798–11801

Pekarek LA, Weichselbaum RR, Beckett MA, Nachman J, Schreiber H 1993 Footprinting of individual tumours and their variants by constitutive cytokine expression patterns. Cancer Res 53:1978–1981

Pekarek LA, Starr BA, Toledano AY, Schreiber H 1995 Inhibition of tumour growth by elimination of granulocytes. J Exp Med 181:435–440

Strasly M, Cavallo F, Geuna M et al 2001 IL-12 inhibition of endothelial cell functions and angiogenesis depends on lymphocyte-endothelial cell cross-talk. J Immunol 166:3890–3899

DISCUSSION

Strieter: One of the reasons that IL12 appears to have failed in the clinic is because it was systemically administered and obviously has significant toxic effects. You showed a number of factors that IL12 seems to be inducing. Do you think any of those potential downstream mediators might be mediating some, if not most, of the effect of IL12? Perhaps one of them could be used and this may avoid the systemic side effects of IL12.

Forni: Potentially, yes. IFNγ is a major mediator induced by IL12. In our hands, however, it has never given the same results as IL12. This is an embarrassing situation that may be linked to the appropriate compartmentalization of the production of downstream cytokines induced by IL12. Systemic administration of IFNγ effectively inhibits tumour progression. The immune reaction it activates, however, is not comparable with that activated by IL12.

Blankenstein: I have a comment concerning the local cytokines. It is well established that tumours transfected to secrete certain cytokines are rejected. If the mice are challenged with the parental tumour a couple of weeks later, this tumour is rejected again, commonly termed tumour immunity. This effect is often attributed to the effect of the transfected cytokine, but from today's perspective this appears problematic. What is done in these experiments is that mice are injected with viable tumour cells, which proliferate before they are rejected. This multiplies the amount of antigen to which the host is exposed. Then, if the same tumour is rejected again this is simply due to the initial exposure of a larger amount of tumour cells to the immune system. What one also observed is that if the same tumour cells expressing the cytokine are irradiated before injection into mice so that they can not proliferate, then the vaccine efficacy drops, even though the cytokine is still produced. This suggests that the amount of antigen is important. I would say today that the local cytokine is primarily modulating the local stroma, which prevents rapid tumour burden. This then allows the CD8 effectors to be activated, but the cytokine produced by the tumour does not necessarily contribute to the generation of systemic tumour immunity. GM-CSF may be an exception.

Forni: I agree with most of what you say. In one experiment we used both a tumour engineered to release cytokine and the same tumour expressing a suicide gene. These tumours grew and then regressed. Both the regression due to cytokine release and the activation of suicide genes significantly increased tumour immunogenicity (Consalvo et al 1995). The prime aspect of achieving tumour rejection by means of cytokine-induced reactions is that one can determine which kind of memory mechanism is elicited. A reaction activated by IL4 will give you a high antibody response but a poor T cell-dependent cytotoxic response. Activation by IFNγ or IL2, on the other hand, results in a much stronger T-cell-dependent cytotoxic response than that activated by the suicide gene. In principle, induction of tumour growth followed by regression is probably the best way to elicit a powerful immunity. This, of course, is unthinkable in humans. Cytokines present at the tumour rejection site skew the memory mechanisms towards Th1 or Th2.

Blankenstein: Do you mean the cytokine that you express in the tumour?

Forni: Yes, or that you inject locally.

Blankenstein: I agree with regard to GM-CSF, because irradiated GM-CSF-producing tumour cells retain their ability to induce increased tumour immunity, perhaps because GM-CSF activates dendritic cells. For other cytokines such as IL4 it has to be reinvestigated whether tumour rejection of the cytokine-producing tumour cells is due to inhibition of stroma formation and/or immune stimulation.

Pollard: I have a question associated with that. In the MMTV HER-2/neu experimental situation, when you have a delay caused by multiple injections of IL12, if you tolerize the mice to HER-2/neu (this is a rat protein expressed only in the mammary epithelium) by neonatal injection, does the IL12 then cause tumour delay? Is this being seen as a foreign tumour because of HER-2/neu, or do other tumour antigens come out on those tumours which allow the IL12 to have its effects?

Forni: The HER-2/neu antigen is already markedly expressed at 3 weeks of age in the rudimentary mammary glands. I cannot prove that these transgenic mice are fully tolerant. They do, however, hyperexpress HER-2/neu at a very early age. By inducing a sustained inflammatory reaction we induce some form of autoimmunity. This, at all events, is my interpretation. I have never tried to render these mice tolerant by specifically injecting soluble proteins.

Oppenheim: It is unfortunate that Steve Rosenberg isn't here, but let me try to do justice to the work he published recently in *Science* (Dudley et al 2002). He starts off with a review of many of his attempts over the last 20 years to use IL2, LAK cells, cloned tumour-infiltrating T cells conditioned and grown *in vitro* with IL2 and transferred adoptively, and so on. Basically what he achieved was a 10–20% temporary remission in a number of patients with a variety of tumours. He recently had much better results which may represent a breakthrough. He has 13 patients who he has followed for three years who are at end-stage disease with melanoma. They have failed all kinds of treatments. He varied the theme somewhat, treating the patients with fluderabine and cytoxan a week before adoptive transfer of cells that had been grown out and cloned against melanoma antigens. The treatment of the patients with fluderabine and cytoxan doesn't touch the tumour, but it kills the patients' lymphocytes. This was done to make room for the cultured autologous lymphocytes, which he transferred adoptively. This consisted of a mixture of CD4 and CD8 cells he had grown in response to the tumour. Previously he had used purified CD8s and they were not effective. He had tried anti-CTLA4 and this resulted in an autoimmune response in one of the patients. With this new approach six out of 13 of the end-stage patients went into complete remissions, and another four had partial remissions. He saw continuing tumour regression for up to six months after the treatments. Several of the patients developed what looked like leukaemia, with enlarged lymphoid cells up to 28 000 per ml. When he added these circulating lymphoblasts to the tumour, they were cytotoxic. What is happening here? Immunologists would agree that administering CD4 helper cells together with CD8 cells is helpful. The 'making room' business is based on the idea that the stromal cells express Flt3 and other stem cell factors on their surface that help the transferred lymphocytes proliferate, and support this new cell population. But the other possibility is that this procedure is also killing tolerant or $CD4^+CD25^+$ anergic cells. It is well accepted that we are dealing with large tumour loads which are shedding enormous amounts of antigen, and that this results in paralysing the immune system. Basically what Dr Rosenberg is doing operationally looks very interesting, and in terms of mechanistic ways of thinking about this it suggests that by suppressing the host's immune system, he is also countering the anergy and paralysis of the system. His approach may therefore have overcome this major problem that can block effective immune responses by tumour patients.

Blankenstein: Certainly, the rationale of this approach is easy to follow. However, it should be emphasized that the idea is 20 years old. Phil Greenberg showed two decades ago that adoptively transferred T cells reject tumours only when the mouse is pre-treated with cyclophosphamide. There are two possible explanations, which you correctly pointed out. Either cyclophosphamide treatment allows homeostatic proliferation of the adoptively transferred T cells or suppressor cells are eliminated, or both. There is a third possibility, which is that cyclophosphamide modulates the tumour stroma. I would also like to point out that the numbers of complete and partial regressions were slightly different.

Oppenheim: It is much better than he has ever achieved before. What he is doing looks very promising.

Pollard: 20 years of research and millions of dollars has cured two patients.

Forni: My point is that inflammation can be a very powerful way to inhibit a proliferating tumour.

Gallimore: One thing worth considering is that regulatory T cells may not work very well under some proinflammatory conditions. For example, there is *in vitro* evidence that they don't work very well in the presence of IL12. This would be an additional explanation for why IL12 provides a good adjuvant effect. This could be the situation with many other cytokines. The inflammatory milieu could therefore have quite a strong effect on the impact of regulatory cells.

Gordon: Guido Forni, can I ask you about the neutrophils? You mentioned them as being significant. Do you think they are contributing to the destruction of the tumour?

Forni: We have plenty of evidence that neutrophils kill tumour cells *in vitro*, together with morphological evidence that the eosinophils and neutrophils intermingled with tumour cells induce massive tumour necrosis.

Gordon: Is this in the absence of antibody?

Forni: Yes. Rejection of tumours engineered to release cytokines occurs within 72–96 h after challenge. This is not long enough to allow a significant production of antibodies.

Gordon: Complement could play a role.

Forni: Yes.

Pollard: Have you seen the same sort of neutrophil infiltration in the ErbB2 tumours?

Forni: Yes, but it is not as massive as when we have local cytokine production. In HER-2/neu transgenic mice IL12 administration is systemic, and this made the situation a little different.

Pollard: Once tissues start to go, they go enormously quickly. Neutrophils are usually the mechanism. Another area that we work in is the placenta, and when we see a resorption within a few hours the placenta has gone. This is almost completely mediated by neutrophils.

Ristimäki: In human tumours we don't usually see neutrophils. We mainly see mononuclear cell infiltration, unless there are foci of necrosis or ulceration. Is this somewhat model dependent? You have a rapidly expanding growth of tumour in your model, and it could be this that attracts the neutrophils. There may also be hypoxia or some subtle necrosis that attracts them.

Forni: I don't think this is the case. Following systemic administration of IL2 we also see a massive neutrophil response in the blood that proceeds in step with tumour shrinkage.

Rollins: Another potential explanation might be the difference between neutrophil attractants in mice versus humans. For example, MIP-1α is a good solid CC chemokine that usually only attracts mononuclear cells in humans, but in mice it is one of the primary neutrophil chemoattractants.

Forni: In any case, when IL2 is administered to mice in which neutrophils have been eliminated by antibody treatment, tumour rejection doesn't take place any longer. So neutrophils appear to be required for the tumour regression activated by local cytokines.

Balkwill: It may be worth considering the leukocytes that are normally in human tumours, and perhaps the mouse tumours as well. The consensus seems to be that in human epithelial tumours we primarily find CD8 lymphocytes and macrophages but very few CD4 cells. This is certainly our experience. We rarely find NK cells or neutrophils in epithelial ovarian cancer. This is the repertoire we have to deal with. Perhaps one of the reasons for the recent success of Rosenberg in treating melanoma patients with lymphokine activated killer cells was that he was adding back CD4 cells, for instance, that are just not present in the tumour, and also ablating some of the immunosuppressive cells that were in the tumour (Dudley et al 2002).

Pollard: That is a good point. When the tumours start to go, they go very quickly. This is when the neutrophils are recruited. Under the normal situation that we see when the tumours come into the clinic, we don't see these types of cells as being particularly abundant.

Balkwill: Even though some of the tumours express neutrophil-attracting chemokines.

Blankenstein: In Guido Forni's case, neutrophils contribute to the rejection of the tumour. There is a paper published a few years ago by Hans Schreiber describing that depletion of granulocytes leads to tumour rejection (Pekarek et al 1995). This tumour apparently needed the granulocytes to grow. I could give similar examples for macrophages. The activation state of the immune cells seems to be important.

Gordon: Some of the antibodies used by people to deplete neutrophils are not specific for neutrophils.

Blankenstein: It was the same antibody that Guido Forni used and which recognizes the Gr1 antigen.

Gordon: That antigen hasn't been very well characterized.

Pollard: I have a general question. In *Listeria* infection, where there is a very strong recruitment early on, induced by macrophages producing neutrophil chemoattractants such as IL8, there is then a burst of IL10. Kaufmann has shown (and we have confirmed) that this is required for the neutrophils to stop coming in. I wonder if it is the macrophage-produced IL10 that is blocking the neutrophils coming into tumours under normal circumstances. Is this IL10 a normal feature of many tumours?

Mantovani: In the tumours that we have looked at, in mouse transplanted tumours and also in human ovarian cancer, IL10 is constitutively expressed by macrophages *in situ*. There are a number of tumours that produce IL10. It is a pretty general feature.

Strieter: The same thing is true in lung cancer.

Pollard: It might be interesting to do a breast cancer model in an IL10 knockout. Has this been done?

Caux: We have used anti-IL10 on mouse transplantable tumours. There is a clear effect of anti-IL10 in combination with a CPG sequence that activates dendritic cells and results in tumour rejection (see Vicari et al 2003, this volume).

References

Consalvo M, Mullen CA, Modesti A et al 1995 5-Fluorocytosine-induced eradication of murine adenocarcinomas engineered to express the cytosine deaminase suicide gene requires host immune competence and leaves an efficient memory. J Immunol 154:5302–5312
Dudley ME, Wunderlich JR, Robbins PF et al 2002 Cancer regression and autoimmunity in patients after clonal repopulation with antitumor lymphocytes. Science 298:850–854

Pekarek LA, Starr BA, Toledano AY, Schreiber H 1995 Inhibition of tumor growth by elimination of granulocytes. J Exp Med 181:435–440
Vicari AP, Vanbervliet B, Massacrier C et al 2004 *In vivo* manipulation of DC migration and activation to elicit anti-tumour immunity. In: Cancer and inflammation. Wiley, Chichester (Novartis Found Symp 256) p 241–258

General discussion II

Mantovani: I wanted to try to bring together the work we have discussed on rheumatoid arthritis (RA) and cancer. We have heard a lot about chemokines in cancer. They interact with seven-transmembrane-domain receptors, and they activate a complex signalling cascade, which eventually leads to phosphatidylinositol-3-kinase (PI3K)γ activation and AKT activation. There is genetic evidence that p10 is a target for inactivation. If one thinks of molecular targets, the knockout mice for the γ isoform indicate that this may be a target for blocking excessive recruitment, and possibly more important things. On the other hand, the lesson that I am trying to take home from Fionula Brennan's data is that the situation is more complicated. I should mention that we have data on the PI3Kγ knockout mice. These have defective capacity to mount certain inflammatory and immune responses. But if we have to learn from *bona fide* inflammatory disorders the situation is more complicated. PI3K would not be a target in RA.

Brennan: Absolutely.

Mantovani: What are the isoforms?

Brennan: We haven't looked at that. Nor have we looked at constitutively active P110. We are looking very much in the macrophages and at production of tumour necrosis factor (TNF), although there might be evidence that the fibroblast proliferation and hyperplasia could be inhibited by blocking PI3K. In an inflammatory environment, several things are going on. PI3K activity is intrinsic to many cell types, but when we focus on a proinflammatory cytokine, TNF, which has been demonstrated in the clinic to be a therapeutic target, we know that blocking PI3K using these drugs that are available, super-induces it. This was unexpected.

Mantovani: Has anyone tried inducing RA in knockout mice for the various PI3K isoforms?

Brennan: No.

Oppenheim: If one thinks about things in a very simplistic manner, autoimmunity represents too much immunity. Cancer represents escape from immunity. You would predict that in autoimmune situations there would be less cancer, despite the inflammation. Frankly, I have not seen any good epidemiological data on any of those issues, but I think of them as mirror images, in a sense.

Richmond: There are good data that immunosuppressed patients have a much stronger tendency to develop tumours.

Blankenstein: We have to be careful here. These are almost always virus-associated tumours. Here we have a completely different situation. It is the response to viruses, not the tumour, which is suppressed.

Pollard: Most of those tumours are not adenocarcinomas in immunosuppressed patients, whereas 95% of normal human tumours are adenocarcinomas.

Rollins: This morning we heard about the increased incidence of adenocarcinomas in inflammatory bowel disease.

Brennan: That probably has an infectious origin.

Harris: What about coeliac disease, where patients get lymphomas?

Ristimäki: RA patients have less gastrointestinal (GI) tract cancer. We don't know whether this is due to the treatment (namely non-steroidal anti-inflammatory drugs, NSAIDs), or the disease (immunological alterations), or both.

Thun: There have been large linkage studies in Sweden and Finland of patients with RA. They were conducted because of the expectation that there would be increased gastric cancer from all of the NSAIDs used to treat patients with rheumatoid arthritis. The principal findings of these studies were a decrease in gastric and colon cancer, and an increase in non-Hodgkin's lymphoma. This was postulated to be the one cancer related to the underlying immunological disorders of RA. In these populations there are conflicting things going on. There are all the effects of the treatment plus the effects of the disease. It is hard to disentangle.

Feldmann: Were these studies age corrected?

Thun: Yes.

Gordon: In your cell–cell interaction model, what are the surface molecules on the T cells and monocytes that you think may modulate this differential activation?

Brennan: We don't know.

Gordon: Presumably you have looked quite hard.

Brennan: It is hard to do those experiments. Other people have done it using blocking antibodies. But we get several answers. If you block the interaction of two cells you will also block the signalling event. But this doesn't demonstrate that it is the molecule involved in the signal induction. It is quite hard to induce TNF production by cross-linking receptors on monocytes. In fact, the only receptor which we find induces TNF is CD45, which occupies about 8% of the cell surface membrane. In the literature there is a lot of background endotoxin in the experiments.

Strieter: I have a question regarding the TNF experiment where there was PI3K inhibition. You saw corresponding superinduction of TNF and a down-regulation of interleukin (IL)10. Is this response responsive to IL10? In other words, if you come back in and reconstitute IL10, would you shut off the superinduction of TNF?

Brennan: No. We still see the effects if we block the endogenous TNF or IL10.

Richmond: In the presence of the PI3K inhibitor, was TNF still active?

Brennan: In the presence of PI3K inhibitor the TNF was super-induced.

Richmond: But is it active on its receptor? In cells that have PI3K blocked, what is the response to TNFα?

Brennan: We haven't looked at that. We would expect that the signalling through the receptor is blocked.

Strieter: In a bioassay you would probably see more death and more TNF bioactive. The biossays for TNF-induced cytotoxicity are probably dependent on PI3K. If this was inhibited they would probably increase the rate of apoptosis. This may not be directly related to the biological activity of TNF.

Ristimäki: With respect to the immune response and cancer, you could also ask about the role of the immune system in treating cancer. We have heard in two papers about HER-2/neu receptor model. There exists an effective antibody against HER-2, and also kinase inhibitors of HER-receptor pathways. Could you use these to examine the contribution of the immune system with respect to treatment, by comparing antibody treatments with kinase inhibitor treatment? Are they as effective, or is one more effective than the other? Does this tell us anything about the immune system working for us in treating cancer?

Forni: This would be possible. We are now immunizing mice in which distinct genes involved in immunological functions are knocked out in order to assess the role of the cell-mediated and the antibody-mediated mechanism in the inhibition of HER-2/neu tumours. Antibody can certainly play a major role by down-regulating the presence of the HER-2/neu protein product from the cell membrane and blocking proliferation. Antibodies alone, however, cannot block the tumour. This is one of the few examples of cooperation between T-cell and humoral reactivity.

Rollins: Is Herceptin an activating antibody as well? If any of these antibodies also lead to some downstream signalling, you will see effects beyond what would be seen with the immune responses. You may be looking at partially non-overlapping Venn diagrams in terms of responses.

Pollard: My understanding of Herceptin is that no one knows how it works, and it doesn't work in the way commonly claimed.

Forni: It works in many ways. In animals, it immediately causes the capping and internalization of the HER-2/neu receptor. The cells then become HER-2/neu negative. Removal of HER-2/neu receptors from the cell membrane is associated with a marked decrease of cell proliferation. We have tumours that are HER-2/neu negative and proliferate very little (Nanni et al 2000). Inhibition of proliferation appears to be the main task of antibody in our mouse system. I have not yet made up my mind about the importance of the killing mechanism it mediates.

Richmond: I have a question relating to Joe Oppenheim's introduction about Steve Rosenberg's trial. What antigen was he using to activate those T cells?

Oppenheim: He was relying on the tumour to provide the antigen.

Richmond: Did he preactivate the cells at all once he put them back in?

Oppenheim: He used all kinds of antigen from the tumour to grow the T cell clones. What actually happened is that sometimes the T cell clones that grew out were different from what he actually initially stimulated them with. They were directed against tumour antigens that were different from what he began with. I forgot to mention that he also repeatedly treated the patients with IL2 after the administration of the adoptively transferred cells. I presume that this was to keep them boosted. He relied on *in vivo* antigen to continue the boosting of the cloned T cells.

Richmond: So he didn't take the patients' own tumour cells.

Oppenheim: I don't think so.

Gordon: Marc Feldmann, you asked what we had learned from the use of TNF in the rheumatoid model. Can you extrapolate from that and say what we should avoid doing in terms of attempts to manipulate tumour responses?

Feldmann: The striking thing that we have learned from blocking TNF in rheumatoid patients is that a single molecule has an enormous number of possible effects. We have a lot of data on TNF blockade, but we are not sure how much these effects would be mimicked by the blocking of other cytokines. I have just written a review where we have compared the blockade of a range of cytokines. The reality is that if you accept the data people are showing, without looking critically, blocking IL6, IL15 and IL1 would give the same results. If these data are all correct — which I am not sure of — it may be that the common denominator is leukocyte trafficking. That is, if we block any one major cytokine we have an impact on trafficking. This is a complicated process which is probably also 'catalytic'. A lot of activated cells in one site make a lot of molecules and suck a lot more in. This seems to be a key aspect. It is probably true also for tumours that the trafficking of cells is quite important. The other thing we have learned is that cancer is different. The chronic inflammatory phenotype seems to be a relatively genetically stable process. If you block TNF the patients that do well stay doing well for many years. There isn't an evolution to resistance and this is probably the dominant difference between chronic inflammation and tumours. With tumours there is an unstable genotype and a tremendous selection process. There are reports of mutations in a synovium. Firestein described p53 mutations and Stephan Gay has described PTEN mutations (Pap et al 2000, Tak et al 1999). These are probably induced by reactive oxygen and nitrogen species. But the frequency is very low, a few cells in a few patients, and they are not selected for. These are almost certainly epiphenomena. Why is it that in one type of chronic disease there is no selection at a rapid rate, whereas in cancer there is?

Gordon: One of the models mentioned by Michael Thun is quite intriguing. It does seem to be a potent factor in promoting mesothelioma.

Thun: Although asbestos is a strong risk factor for mesothelioma, mesothelioma is rare even among asbestos-exposed populations. Most of the lung cancers caused by asbestos are squamous cell carcinomas, and most result from the combination of smoking and asbestos. A key issue of debate is, to what extent does the asbestos act as a mechanical conveyor of the carcinogens into cells? I haven't heard any discussion of the fibrosis being a promoter for squamous cell carcinomas. When I presented a preview of my paper to our group, the issue was raised that idiopathic pulmonary fibrosis is only weakly associated with increased lung cancer. One might expect a stronger association if the inflammation hypothesis were real. However, my understanding of idiopathic pulmonary fibrosis is that the inflammation occurs diffusely rather than being localized to any specific part of the lung. This differs from the highly localized and sustained inflammation in Barrett's oesophagus.

Gordon: How about the schistosome model? You mentioned bilharzia.

Thun: That is a very chronic inflammation, and very localized. The patients are either not clearing the schistosome or they are constantly getting reinfected because they are agricultural workers in contaminated water. They meet the requirement of having a chronic infection plus it is very localized. The schistosomes take up residence in the terminal ureters, and they lay their eggs such that the biggest exposure is right where the ureters empty into the bladder.

Gordon: Is the incidence of bladder cancer quite high in terms of the overall risk in the populations where bilharzia is endemic?

Thun: It is extremely high. In fact, bladder cancer comprises a remarkable fraction of the incident cancers in Egypt.

Ristimäki: It is a relatively rare type of urinary bladder cancer (squamous cell type) in the Western world.

Thun: It is actually intriguing that the histological type of bladder cancer associated with schistosomiasis is squamous cell carcinoma, not transitional cell bladder carcinoma that is the principal type in other countries.

Ristimäki: I think it is because the schistosomes cause metaplasia that is then prone to form dysplasia and cancer. In lungs it is the respiratory epithelium that turns to squamous metaplastic epithelium that is prone to transformation.

Thun: Several occupational bladder carcinogens also cause progressive atypia. I don't know whether the cancers that these give rise to are squamous or transitional cell.

Harris: It is transitional, but that is a chemical carcinogen.

Gordon: The other factor with schistosomiasis is the strong Th2 response, rather than the Th1 response seen in tuberculosis.

D'Incalci: Since there are many PI3K inhibitors in early clinical development, what do immunologists predict in terms of positive and negative effects on the immune system?

Mantovani: For selective inhibitors of the gamma isoform of PI3K, the prediction is that there would be lowered recruitment of macrophages. This may be a positive effect. There would be a reduction of delayed-type hypersensitivity and contact hypersensitivity. There would also be a reduction of specific immunity. This is what the knockout mice tell us. I don't know whether the knockouts of the other PI3K isoforms have been studied accurately.

Brennan: Interestingly, in the contact-mediated model, if we activate the T cells conventionally with a mimic for antigen, this is a PI3K dependent event. When we activate the T cells with anti-CD3, introduce them to monocytes and look for TNF production, PI3K inhibitors down-regulate this. We see that the superinduction of TNF is clearly associated with the inflammatory response as opposed to an immune response.

Mantovani: Which tumours are they targeted to?

D'Incalci: In ovarian cancer there is amplification of the PI3Kγ isoform gene.

References

Nanni P, Pupa SM, Nicoletti G et al 2000 p185(neu) protein is required for tumor and anchorage-independent growth, not for cell proliferation of transgenic mammary carcinoma. Int J Cancer 87:186–194

Pap T, Franz JK, Hummel KM, Jeisy E, Gay R, Gay S 2000 Activation of synovial fibroblasts in rheumatoid arthritis: lack of Expression of the tumour suppressor PTEN at sites of invasive growth and destruction. Arthritis Res 2:59–64 (free full-text article at *http://arthritis-research.com/content/2/1/59*)

Tak PP, Smeets TJ, Boyle DL et al 1999 p53 overexpression in synovial tissue from patients with early and longstanding rheumatoid arthritis compared with patients with reactive arthritis and osteoarthritis. Arthritis Rheum 42:948–953

Lymphangiogenesis and tumour metastasis

Jean-Christophe Tille, Riccardo Nisato and Michael S. Pepper[1]

Department of Morphology, University Medical Center, 1 rue Michel Servet, 1211 Geneva 4, Switzerland

Abstract. The lymphatic system serves to collect and transport interstitial fluid (lymph) within tissues, and plays an important role in the immune response. The lymphatic system also constitutes one of most important pathways of tumour dissemination. In several human cancers, increased expression in primary tumours of a new member of the vascular endothelial growth factor (VEGF) family, namely VEGF-C, is correlated with regional lymph node metastasis. Experimental studies using transgenic mice overexpressing VEGF-C or xenotransplantation of VEGF-C-expressing tumour cells into immunodeficient mice have demonstrated a role for VEGF-C in tumour lymphangiogenesis and the subsequent formation of lymph node metastasis. However, there is at present very little evidence for lymphangiogenesis in human tumours, which is at variance with the data obtained in animal models. Nonetheless, the striking correlation between levels of VEGF-C in primary tumours and lymph node metastases exists. This suggests that VEGF-C may activate pre-existing lymphatics which then become actively involved in tumour cell chemotaxis, intralymphatic intravasation and distal dissemination. The role of VEGF-C in human tumour metastasis is therefore likely to involve lymphangiogenesis as well as its capacity to induce activation of pre-existing lymphatic endothelium.

2004 Cancer and inflammation. Wiley, Chichester (Novartis Foundation Symposium 256) p 112–136

The lymphatic system serves to collect and transport extravasated protein-rich fluid (lymph), macromolecules and cells of the immune system within organs and tissues. This extensive drainage network is lined by a continuous layer of endothelial cells and is interspaced by lymph nodes. It begins in the tissues as a series of blind-ending capillaries which drain into collecting vessels that in turn

---

[1]This paper was presented at the symposium by Michael S. Pepper to whom correspondence should be addressed.

return lymph to the systemic blood circulation via the thoracic duct. When lymphatic circulation is compromised due to mechanical obstruction, lymphoedema ensues. In addition to providing a means for interstitial fluid return to the circulation, the lymphatic system plays an important role in the immune response by directing the circulation of lymphocytes and antigen presenting cells. With regard to pathology, the lymphatic system constitutes one of most important pathways for tumour dissemination (Pepper 2001, Stacker et al 2002).

Despite the longstanding recognition of the involvement of the lymphatic system in many clinical settings, formal experimental demonstration of its involvement (including lymphangiogenesis) in the pathogenesis of lymphoedema or tumour cell dissemination in experimental models has until recently been lacking. This was due to the lack of lymphatic-specific markers as well as identification of lymphangiogenic growth factors and their receptors. However, much progress has been made in these areas in the past decade. It is now also possible to culture pure populations of blood and lymphatic vascular endothelial cells using the molecular tools which are available. In this respect, one can truly speak of a renaissance in the field of lymphatic endothelial biology and pathophysiology.

Lymphatic endothelial cell biology: a renaissance

Lymphatic endothelial cell markers

Until the discovery of specific markers, lymphatic vessels were identified following the uptake of dyes such as Evans blue, ferritin, trypan blue or patent blue. The absence of basement membrane components such as laminin, collagen IV and collagen XVIII as well as the lack of PAL-E staining of CD31-positive endothelial cells were also used as lymphatic endothelial-specific criteria. 5'-nucleotidase was also used by some investigators to identify lymphatic endothelium.

In the past few years, a number of relatively specific lymphatic endothelial markers have been discovered. These include vascular endothelial growth factor receptor 3 (VEGFR-3), which is predominantly expressed by lymphatic endothelial cells in normal adult tissues, but which is also expressed by blood vascular endothelial cells in tumours or during wound healing. Other transmembrane proteins such as LYVE-1, podoplanin or desmoplakin have been shown to be reliable lymphatic markers, although none are strictly lymphatic endothelial-specific. In the vascular system, the transcriptional factor Prox-1 appears to be specific for lymphatic endothelium, although it is expressed in other cell types in other tissues (Sleeman et al 2001).

Lymphangiogenic growth factors

Two lymphangiogenic factors have been identified to date, and both belong to the VEGF family. Other molecules which play an important role in the development and maintenance of the lymphatic system include angiopoietin 2 (Gale et al 2002), neuropilin 2 (Yuan et al 2002) and Prox-1 (Wigle & Oliver 1999).

The VEGF family is comprised of several secreted glycoproteins that play a prominent role in the formation of blood and lymphatic vessels (angiogenesis and lymphangiogenesis). The mammalian VEGF family consists of VEGFs -A, -B, -C and -D. VEGF signalling in endothelial cells occurs through three tyrosine kinase receptors (VEGFRs): VEGFRs -1, -2 and -3. VEGFR-1 binds to VEGFs -A and -B, VEGFR-2 binds VEGFs -A, -C and -D whereas VEGFR-3 interacts with VEGFs -C and -D. In adult tissues VEGFRs -1 and -2 are predominantly expressed by vascular endothelial cells and promote endothelial cell proliferation, migration, and angiogenesis. VEGFR-3, which is widely expressed in the early embryonic vasculature, becomes restricted to lymphatic endothelial cells in the later stages of embryogenesis and in the adult. VEGFs -C and -D are secreted as homodimers that undergo extensive proteolytic processing of their N- and C-terminal domains following secretion. The enzymes responsible for processing have not been identified. Processing of VEGFs -C and -D alters their binding affinities for VEGFRs, thereby modulating the biological effects of these growth factors. The secreted 31 kDa form of VEGF-C predominantly activates VEGFR-3 whereas the mature fully processed 21 kDa form, corresponding to the central VEGF homology domain (VHD), activates both VEGFRs -2 and -3.

VEGFs -C and -D have the dual capacity to induce lymphangiogenesis and angiogenesis. This has been demonstrated in a number of experimental systems including the chick chorioallantoic membrane (CAM), the rabbit cornea assay and transgenic mice (Table 1). Two major strategies have been adopted: some groups have studied the growth factors themselves, while others have focused on VEGFR-3.

Application of recombinant VEGF-C protein to the differentiated avian CAM or the mouse cornea (Kubo et al 2002, Oh et al 1997) as well as application of tumour cells onto the differentiated avian CAM (Papoutsi et al 2000, 2001) (Table 5) leads to the formation of lymphatic capillaries. Lymphatic endothelial cells in these settings were shown to be proliferating and their unusual location suggested that the lymphatics were newly formed. Viral gene delivery of fully-processed VEGF-D in the rat skin induced a significant lymphangiogenic effect together with a (less robust) angiogenic response, whereas viral gene delivery of VEGF-A in the same setting only induced angiogenesis (Byzova et al 2002). In transgenic mice expressing VEGF-D or VEGF-C or a VEGF-C mutant (VEGF-C-156S) that binds VEGFR-3 but not VEGFR-2, VEGFR-3 signalling was shown

TABLE 1 Experimental models of lymphangiogenesis

Growth factors/ receptors	Receptor/growth factor specificity	Lymphatic marker	Animal model	References
VEGF-C	VEGFR-2, VEGFR-3	VEGFR-3	Transgenic expression in mouse skin	Jeltsch et al 1997
		LYVE-1	Viral gene delivery in mouse skin	Enholm et al 2001
		LYVE-1, VEGFR-3	Recombinant protein pellets implanted into mouse cornea	Kubo et al 2002
		VEGFR-3	Application of recombinant protein in CAM assay	Oh et al 1997
		LYVE-1, Podoplanin	Viral gene delivery or skin-specific transgenic expression in Chy mice	Karkkainen et al 2001, Saaristo et al 2002
VEGF-C156S	VEGFR-3	LYVE-1, VEGFR-3	Transgenic expression in mouse skin	Veikkola et al 2001
		LYVE-1, Podoplanin	Viral gene delivery or skin-specific transgenic expression in Chy mice	Karkkainen et al 2001, Saaristo et al 2002
VEGF-D	VEGFR-2, VEGFR-3	VEGFR-3	Viral gene delivery in rat skin	Byzova et al 2002
		LYVE-1, VEGFR-3	Transgenic expression in mouse skin	Veikkola et al 2001
VEGFR-3	VEGF-C, VEGF-D	LYVE-1, VEGFR-3	Transgenic expression of VEGFR-3-Ig in mouse skin	Makinen et al 2001a

to be sufficient to mediate selective lymphangiogenesis without affecting angiogenesis (Enholm et al 2001, Jeltsch et al 1997, Veikkola et al 2001). During early embryogenesis in VEGFR-3-null mice, lack of VEGFR-3 neither affects vasculogenesis nor angiogenesis. Instead, vessel remodelling and maturation are impaired, and this leads to embryonic death at day 9.5, well before the emergence of the lymphatic vasculature (Dumont et al 1998). Selective expression of soluble VEGFR-3-Ig in mouse skin led to a loss of lymphatics during embryogenesis, demonstrating that sequestration of VEGFs -C and -D prevents formation of initial lymphatics without affecting the blood vasculature. Lymphatics appeared spontaneously in older mice, suggesting that other factors can compensate for the loss of VEGF-C- or -D-induced VEGFR-3 signalling in postnatal life (Makinen et al 2001a). Further evidence for the importance of these ligand-receptor interactions for lymphatic activation has come from mice carrying a spontaneous mutation in VEGFR-3 (Chy mice) (Karkkainen et al 2001, Saaristo et al 2002) as well as from transgenic mice expressing soluble VEGFR-3-Ig (Makinen et al 2001a). These mice displayed features of lymphoedema, which could be reversed by VEGF-C or a VEGF-C mutant (VEGF-C-156S) that binds only VEGFR-3.

These reports clearly demonstrate that VEGF-C and/or VEGF-D signalling through VEGFR-3 induces lymphangiogenesis. However, some studies reported that these cytokines are also capable of inducing angiogenesis, an effect that could be mediated by fully processed growth factor that activates blood vascular endothelium via VEGFR-2 (Byzova et al 2002, Veikkola et al 2001). In wound healing (Paavonen et al 2000) or in tumours (Kubo et al 2000), blood vascular endothelial cells up-regulate VEGFR-3. Whether this is involved in signalling which leads to angiogenesis remains to be established.

In summary, it appears from *in vivo* and *in vitro* studies that VEGFR-2 signalling regulates angiogenesis whereas VEGFR-3 signalling mediates lymphangiogenesis. However, the role of VEGFR-2 in lymphangiogenesis remains to be established, as does the role of VEGFR-3 in angiogenesis in certain settings.

Cultured lymphatic endothelial cells

Very few differentially expressed molecules have been identified to date that distinguish lymphatic from blood vascular endothelium, and the extent to which the two cell types are related remains unclear. This scenario has changed very recently with reports which describe the isolation of lymphatic endothelial cells (LECs) from human dermis (Kriehuber et al 2001, Makinen et al 2001b, Podgrabinska et al 2002). Positive and negative selection was employed to isolate LECs and blood endothelial cells (BECs) from mixed populations of dermally derived endothelial cells. LYVE-1, podoplanin, VEGFR-3, CD31 and CD34 were used for cell sorting in the various studies. LECs are LYVE-1$^+$/podoplanin$^+$/VEGFR-3^{high}/CD31$^+$/CD34^{low}, whereas BECs are LYVE-1$^-$/podoplanin$^-$/VEGFR-3$^{low/-}$/CD31$^+$/CD34^{high}. BEC selectively expressed CD44 and the PAL-E-reactive antigen. Both cell types expressed VE-cadherin and von Willebrand factor (vWF). Neither cell type expressed CD45, α smooth muscle actin, cytokeratins or neurofilaments. One of the most striking findings in these studies was the homotypic sorting of LEC and BEC when the two cell types were co-cultured (Kriehuber et al 2001, Makinen et al 2001b).

BECs and LECs express distinct sets of vascular markers and they respond differently to growth factors and extracellular matrix. A comparative analysis of gene expression profiles has been done using microarrays (Petrova et al 2002, Podgrabinska et al 2002). With respect to lymphatic endothelium, classification into functional groups revealed particularly high levels of genes implicated in protein sorting and trafficking, indicating a more active role of lymphatic endothelium in uptake and transport of proteins than had previously been anticipated (Podgrabinska et al 2002). These studies should facilitate the discovery of novel lymphatic vessel markers, and provide a basis for the analysis of the molecular mechanisms that account for the characteristic functions of lymphatic capillaries.

Lymphangiogenic factors and lymphangiogenesis in human tumours

Expression of lymphangiogenic factors in human tumours

The discovery of the lymphangiogenic factors VEGFs -C and -D raised the question as to whether these factors are expressed in human cancers and if so whether this contributes to the ability of tumours to metastasize.

VEGF-C is expressed in one-half of human tumours examined to date, principally in the cytoplasm of tumour cells (Salven et al 1998). Its expression is not homogenous in all cells but is found preferentially at the deepest invasive site (Amioka et al 2002, Furudoi et al 2002). Although less is known about the presence of VEGF-D in human tumours, it has been detected in tumour cells in melanoma and colorectal carcinoma (Achen et al 2001, White et al 2002).

Expression studies have allowed a direct comparison between levels of VEGFs -C and -D with clinicopathological features related to the ability of primary tumours to spread (i.e. lymphatic vessel invasion, lymph node involvement and disease-free survival). The majority of these studies revealed a significant correlation between VEGF-C levels in the primary tumour and clinical parameters of tumour spread (Table 2) including lymphatic vessel invasion (LVI) or lymph node metastasis.

TABLE 2   Relationship between VEGF-C levels in primary human tumours and lymph node metastases

Tumour type	VEGF-C detection	Relationship between VEGF-C and metastases	Comment	References
Thyroid carcinoma	IHC	LN		Tanaka et al 2002a
	RT-PCR	LN		Tanaka et al 2002b
	RT-PCR, IHC	LN		Bunone et al 1999
Oesophageal SCCs	IHC	LVI, LN	Increased MVD	Kitadai et al 2001
Gastric carcinoma	IHC	LVI	Poor DFS 5 years	Ichikura et al 2001
	RT-PCR, IHC,WB	LVI, LN	Poor DFS 5 years	Yonemura et al 1999
	IHC	LVI, none LN		Kabashima et al 2001
	IHC	LVI, LN	Poor DFS 6 years	Takahashi et al 2002
	IHC	LVI, LN	Increased MVD	Amioka et al 2002
Breast carcinoma	IHC	None LN	Poor DFS 8 years	Yang et al 2002
	RT-PCR, IHC	LVI	Poor DFS 5 years	Kinoshita et al 2001
	RT-PCR, IHC	None LN		Gunningham et al 2000
	RT-PCR	LN		Kurebayashi et al 1999
	IHC	None LN	Inflammation with increased LVD. Correlation between LVI and LN	Schoppmann et al 2002
Cervical carcinoma	RT-PCR	LN	Poor DFS 5 years	Hashimoto et al 2001
	IHC	LN	Increased MVD, poor DFS 8 years	Ueda et al 2002

	Method		Findings	Reference
Lung carcinoma	IHC	LVI, LN		Kajita et al 2001
	IHC	LN		Ohta et al 2000
	RT-PCR	LVI		Niki et al 2000
Mesothelioma	RT-PCR 5'-Nase	LVI, none LN	Poor DFS 3 years	Ohta et al 1999
Pancreatic carcinoma	IHC	LVI, LN	No relationship to DFS 5 years	Tang et al 2001
Endometrial carcinoma	IHC	LVI, LN		Hirai et al 2001
Neuroblastoma	RT-PCR	None LN		Komuro et al 2001
Prostatic carcinoma	ISH	LN	Poor DFS 5–10 years	Tsurusaki et al 1999
Colorectal carcinoma	RT-PCR	None LN		George et al 2001
	RT-PCR, IHC	LVI, LN		Akagi et al 2000
Head and Neck SCC	IHC	LVI, LN	Increased MVD, poor DFS 5 years	Furudoi et al 2002
	RT-PCR, WB	LN		O-Charoenrat et al 2001
	RT-PCR		No relationship between VEGF-C levels and intratumoural lymphatic vessel proliferation	Beasley et al 2002

DFS, disease-free survival; IHC, immunohistochemistry; ISH, *in situ* hybridization; LN, lymph node; LVD, lymphatic vessel density; LVI, lymphatic vessel invasion; MVD, microvascular density; RT-PCR, reverse transcriptase polymerase chain reaction; SCC, squamous cell carcinoma. WB, western blot; 5'-Nase, 5'-nucleotidase.

Some studies have suggested that expression of VEGF-D in human tumours (Table 3) is reduced relative to normal tissue (George et al 2001, O-Charoenrat et al 2001). The relationship between expression of VEGFs -C and -D in human tumours remains to be clarified (Niki et al 2000, O-Charoenrat et al 2001). Another avenue thus far unexplored is the possible role of VEGFs -C and -D as survival factors for tumour cells in certain cancer types (Fielder et al 1997, Orpana & Salven 2002).

Lymphatic vessels in human tumours

The discovery of the lymphatic endothelial markers LYVE-1, podoplanin and VEGFR-3 has allowed for the assessment of lymphangiogenesis during tumour progression. Despite the expression of VEGFs -C and -D by human tumour cells, to date no evidence has been found for the existence of functional intratumoural lymphatic vessels (de Waal et al 1997, Padera et al 2002, Schoppmann et al 2002). In head and neck carcinoma, intratumoural lymphatic vessels have been observed (Beasley et al 2002), but this could have resulted from co-option of pre-existing lymphatics into the rapidly expanding tumour, particularly since this region of the body contains a high density of lymphatic vessels. The fact that the endothelial cells of these vessels were shown to be proliferating does not imply that lymphangiogenesis was occurring, since pre-existing lymphatic endothelium could conceivably have been induced to proliferate in the presence of tumour cell-derived VEGF-C or -D.

On the other hand, dilated lymphatic vessels, in which endothelial proliferation is often observed, are very frequently present at the border of many tumour types (Beasley et al 2002, Niki et al 2001, Padera et al 2002). Expression studies have allowed a direct comparison between lymphatic vessel markers with clinicopathological parameters (Table 4). In several studies, a strong correlation was found between the presence of lymphatic markers (e.g. podoplanin, VEGFR-3) and lymph node metastasis. Tumour cells are frequently found in juxtatumoural lymphatics in human tumours and many studies have reported a strong correlation between lymphatic vessel invasion and lymph node involvement (Table 4).

Existing lymphatic vessels provide a readily accessible avenue for tumour cell dissemination. Tumour spread via the lymphatic vascular bed may therefore be facilitated by the high intrinsic lymphatic density in the tissue in which the tumour arises. The relative contribution of pre-existing versus new lymphatic vessels to lymphogenous metastasis is poorly understood. With respect to lymphatic vessel density (LVD), a correlation exists between the number of tumour-associated lymphatics and the presence of lymph node metastases for a given tumour type (Table 4). However, a comparison of LVD in the tumour

TABLE 3 Relationship between VEGF-D levels in primary human tumours and lymph node metastases

Tumour type	VEGF-D detection	Relationship between VEGF-D and metastases	Comment	References
Breast carcinoma	RT-PCR	None LN	Inflammatory response	Kurebayashi et al 1999
Thyroid carcinoma	RT-PCR	None LN		Tanaka et al 2002b
Lung carcinoma	RT-PCR	LN	VEGF-D level in tumour lower than in normal tissue	Niki et al 2000
Head and Neck SCC	RT-PCR WB	None LN	VEGF-D level in tumour lower than in normal tissue	O-Charoenrat et al 2001
Colorectal carcinoma	RT-PCR	None LN	VEGF-D level in tumour lower than in normal tissue	George et al 2001
	IHC	LN	Poor DFS 7 years	White et al 2002

IHC, immunohistochemistry; RT-PCR, reverse transcriptase polymerase chain reaction; WB, western blot; LN, lymph node; DFS, disease-free survival; SCC, squamous cell carcinoma.

TABLE 4 Relationship between lymphatic density in primary human tumours and lymph node metastases

Tumour type	Lymphatic detection	Correlation between LVD and metastasis	LVD in normal tissue and tumour	Comment	References
Lung carcinoma	IHC (VEGFR-3)	LN		Ki67 in lymphatic vessels	Niki et al 2001
Melanoma	IHC (CD31⁺/PAL-E⁻)	n.d.	LVD comparable to normal skin		de Waal et al 1997
Endometrial carcinoma	IHC (VEGFR-3)	LN		Poor DFS 5 years	Yokoyama et al 2000
Gastric carcinoma	IHC (VEGFR-3)	LVI, LN	Trend to increased LVD in carcinoma (p < 0.067)	Correlation between VEGF-C level and VEGFR-3 level in tumour	Yonemura et al 2001
Tongue SCC	IHC (VEGFR-3)	LN			Okamoto et al 2002
Head and Neck SCC	IHC (LYVE-1/CD34)	LN (only in oropharyngeal carcinoma)		Detection of intratumoural lymphatic vessels	Beasley et al 2002
Cervical carcinoma	IHC (podoplanin)	None LN		Correlation between LVI and LN	Schoppmann et al 2001b
	IHC (podoplanin)	None LN		Correlation between LVI and LN with poor DFS 10 years	Birner et al 2001
	IHC (podoplanin)	n.d.	Trend to increased LVD in carcinoma (p < 0.078)		Schoppmann et al 2002
Colorectal carcinoma	IHC (VEGFR-3)	n.d.	Increased LVD in carcinoma (p < 0.001)		White et al 2002
Breast carcinoma	IHC (VEGFR-3)	LN		Correlation between LVI and LN but not LVD and LN	Nathanson et al 2000
	IHC (podoplanin)	LN			Schoppmann et al 2001a

LVI, lymphatic vessel invasion; LN, lymph node; LVD, lymphatic vessel density; DFS, disease-free survival; SCC, squamous cell carcinoma; IHC, immunohistochemistry; n.d., not determined.

with LVD in the tissue in which it arose has only been undertaken in a limited number of studies. Only a single study (on colorectal carcinoma) has shown that LVD is increased in tumours relative to normal colonic mucosa (White et al 2002). In two studies, although there was a trend towards an increase in tumour-associated LVD, this did not reach statistical significance (Schoppmann et al 2002, Yonemura et al 2001). In one study, melanoma LVD was comparable to LVD in normal skin (de Waal et al 1997). Taken together, these studies provide very limited evidence for the existence of tumour lymphangiogenesis.

It is not known to what extent tumour cell-secreted factors are directly responsible for the formation of the large lymphatic vessels that are detected around human tumours. Inflammatory cells could for example also contribute to lymphatic enlargement (and possibly lymphangiogenesis), particularly since VEGF-C is chemotactic for macrophages (Skobe et al 2001a) and is readily induced by proinflammatory cytokines (Narko et al 1999). A recent study has demonstrated that in breast carcinoma, tumour-associated macrophages express VEGFs -C and -D, thereby suggesting the existence of an additional stimulus for peritumoural lymphatic endothelium (Schoppmann et al 2002). Furthermore, a significant correlation between the tumour inflammatory response and lymphangiogenic factor expression has been observed in breast carcinoma and in cervical carcinoma (Kurebayashi et al 1999, Schoppmann et al 2002).

Experimental models implicating lymphangiogenesis in tumour spread

Tumour spread occurs via both haematogenous and lymphogenous pathways. The discovery of the relatively lymphatic-specific growth factors VEGFs -C and -D and their receptor VEGFR-3 has provided the molecular tools with which to study solid tumour dissemination via lymphatics. Although clinicopathological studies provided a correlation between VEGF-C expression in primary tumours and tumour spread via lymphatics, until recently no direct demonstration of the involvement of VEGF-C or -D had been documented.

On the basis of previous studies which had demonstrated the lymphangiogenic effect of VEGF-C, and knowing that the spread of tumour cells via lymphatic vessels was required for their dissemination to lymph nodes, we asked whether VEGF-C-induced lymphangiogenesis might promote the formation of regional lymph node metastases. To address this question, we established two transgenic mouse lines in which VEGF-C expression, driven by the rat insulin promoter (RIP), was targeted to β cells of the endocrine pancreas. Transgenic RIP-VEGF-C mice develop an extensive network of lymphatics around islets of Langerhans, as shown by a number of criteria including staining with the lymphatic endothelial cell-specific marker, LYVE-1 (Mandriota et al 2001). As a model of tumour

TABLE 5 Experimental models of lymphangiogenesis related to tumour metastasis

Cytokine/ receptor	Transfected tumour cell line	Animal model	Lymphatic marker	Tumour-associated lymphatics			Lymph node metastasis	Reference
				Peritumoural	Intratumoural	Angiogenesis		
VEGF-C endogenous	A375 human melanomas	Avian CAM	Prox-1	Yes Inhibited by VEGFR-3-Ig	Yes Inhibited by VEGFR-3-Ig	Not inhibited by VEGFR-3 Ig	n.d.	Papoutsi et al 2000
VEGF-C endogenous	10AS rat pancreatic carcinoma	Avian CAM	VEGFR-3	Yes	Yes	n.d.	n.d.	Papoutsi et al 2001
VEGF-C over-expression	MDA-MB-435 human melanoma cell line	Nude mice	LYVE-1, VEGFR-3	Yes	Yes	No	Yes	Skobe et al 2001a
VEGF-C over-expression	MeWo human melanoma cell line	Nude mice	LYVE-1, VEGFR-3	Yes	Yes	Yes	n.d.	Skobe et al 2001b
VEGF-C over-expression	MCF7 human breast cancer cells	SCID mice	LYVE-1, VEGFR-3	Yes Inhibited by VEGFR-3-Ig	Yes Inhibited by VEGFR-3-Ig	No	No	Karpanen et al 2001
VEGF-C over-expression	MCF7 human breast cancer	Nude mice	LYVE-1, VEGFR3	Yes	Yes	Yes	Yes	Mattila et al 2002

VEGF-C over-expression	Murine T241 fibrosarcoma or B16-F10 melanoma	Nude mice	LYVE-1, Prox-1	Yes	Yes	Yes	Yes	Padera et al 2002
VEGF-C over-expression	RIP1-Tag2 Pancreatic β cells transgenic mice	RIP1-Tag2 transgenic mice	LYVE-1	Yes	No	No	Yes	Mandriota et al 2001
VEGF-D over-expression	293EBNA cells	SCID mice	LYVE-1	Yes Inhibited by VEGF-D neutralizing antibody	Yes Inhibited by VEGF-D neutralizing antibody	Yes Inhibited by VEGF-D neutralizing antibody	Yes Inhibited by VEGF-D neutralizing antibody	Stacker et al 2001
VEGFR-3-Ig over-expression	LNM35 human cancer cell line expressing endogenous VEGF-C	SCID mice	LYVE-1	Yes Inhibited by VEGFR-3-Ig	Not affected by VEGFR-3-Ig	Yes Inhibited by VEGFR-3-Ig	Yes Inhibited by VEGFR-3-Ig	He et al 2002

n.d., not determined.

progression in the same tissue, we used RIP1-Tag2 mice (Hanahan 1985) in which expression of the SV40 oncogene is driven in islet beta-cells by the rat insulin promoter. These mice predictably and reproducibly develop pancreatic beta-cell tumours which are not metastatic. When RIP-VEGF-C and RIP1-Tag2 mice were crossed, we found that double transgenic mice formed tumours surrounded by well-developed lymphatics, and that this was accompanied by the formation of metastases in regional pancreatic lymph nodes (Mandriota et al 2001). Of importance is the finding that in the same model, tumour cells overexpressing VEGF-A promoted angiogenesis and tumour growth, but did not promote either lymphangiogenesis or the formation of lymph node metastasis (Gannon et al 2002).

Several studies have shown that injection of VEGF-C overexpressing tumour cells into immunocompromised mice promotes the formation of lymphatic-dependent metastases in regional lymph nodes (Table 5). Interestingly, Skobe et al (2000a) showed that VEGF-C induced lymphangiogenesis not only correlated with lymph node metastasis, but also that intra-tumoural lymphangiogenesis correlated with metastatic tumour area in the lung. To date, only one report has implicated VEGF-D in lymphatic-dependent tumour cell dissemination (Stacker et al 2001). When tumour cells overexpressing VEGF-A were assessed in the same model, angiogenesis and tumour growth were increased with no effect on either lymphangiogenesis or the formation of lymph node metastasis.

In some reports, VEGF-C and VEGF-D also promoted a modest angiogenic response as well as tumour growth (Karpanen et al 2001, Skobe et al 2001b, Stacker et al 2001). This may reflect the degree of processing of the growth factor.

Other approaches have targeted VEGFR-3 signalling using tumour cells expressing high levels of VEGF-C (He et al 2002, Karpanen et al 2001). In these models, a soluble form of VEGFR-3 (VEGFR-3-Ig) was an efficient inhibitor of lymphangiogenesis, and this correlated with a marked reduction in lymph node metastasis.

Certain of the experimental studies described above, in which tumour cells overexpressing VEGF-C or -D were implanted into immunocompromised mice, revealed the presence of lymphatics within the tumour. These intra-tumoural lymphatics were essentially located in the outer region of the tumour and appeared to be compressed and destroyed. Two recent reports have demonstrated the absence of functional intra-tumoural lymphatics in tumours induced by sarcoma cells expressing high levels of VEGF-C or in tumour cells transfected with VEGF-C (Leu et al 2000, Padera et al 2002). These reports suggest that VEGF-C and/or VEGF-D signalling through VEGFR-3 leads to lymphangiogenesis, and that lymphatics at the tumour margin are sufficient for lymphatic metastasis.

Conclusion and perspectives

A direct role for VEGF-C in tumour lymphangiogenesis and lymph node metastasis formation has been demonstrated in animal models. Furthermore, increased expression of VEGF-C in human primary tumours correlates with the presence of regional lymph node metastases. However, there is at present very little evidence for lymphangiogenesis around or within human tumours, which is at variance with the data obtained in animal models. Nonetheless, the striking correlation between levels of VEGF-C in primary tumours and lymph node metastases still remains.

As with the blood vascular system, tumour cell dissemination via lymphatics requires their intravasation into lymphatic capillaries. However, very little is known about how this process is regulated, but it is conceivable that lymphatic endothelial cells and tumour cells enter into a reciprocal dialogue. One hypothesis is that VEGF-C may alter the function of pre-existing lymphatics allowing them to become actively involved in tumour cell chemotaxis, lymphatic intravasation and dissemination. Chemokines, with their well-known chemotactic properties, may be the mediators of directional tumour cell migration. VEGF-C may also alter adhesion molecule expression in lymphatic endothelial cells. Formal proof of these hypotheses is still awaited. An alternative hypothesis, namely that tumour cells are passively washed into the blind-ending lymphatic capillaries along with interstitial fluid, has been proposed (Hartveit 1990).

Understanding the molecular and cellular mechanisms of metastasis is essential for the development of new forms of cancer therapy. In pre-clinical studies, molecules which have been shown to be effective in inhibiting tumour lymphangiogenesis and lymph node metastasis include a soluble VEGFR-3–IgG fusion protein and neutralizing anti-VEGF-D antibodies. One group has synthesized and characterized indolinones that differentially block VEGF-C- and -D-induced VEGFR-3 kinase activity but not that of VEGFR-2 (Kirkin et al 2001). These tools provide a glimpse of what could potentially be a novel therapeutic opportunity for the prevention of tumour cell dissemination and metastasis formation.

Acknowledgements

Work in the authors' laboratory was supported by a grant from the Swiss National Science Foundation (no. 3100-064037.00).

References

Achen MG, Williams RA, Minekus MP et al 2001 Localization of vascular endothelial growth factor-D in malignant melanoma suggests a role in tumour angiogenesis. J Pathol 193:147–154

Akagi K, Ikeda Y, Miyazaki M et al 2000 Vascular endothelial growth factor-C (VEGF-C) expression in human colorectal cancer tissues. Br J Cancer 83:887–891

Amioka T, Kitadai Y, Tanaka S et al 2002 Vascular endothelial growth factor-C expression predicts lymph node metastasis of human gastric carcinomas invading the submucosa. Eur J Cancer 38:1413–1419

Beasley NJ, Prevo R, Banerji S et al 2002 Intratumoral lymphangiogenesis and lymph node metastasis in head and neck cancer. Cancer Res 62:1315–1320

Birner P, Obermair A, Schindl M, Kowalski H, Breitenecker G, Oberhuber G 2001 Selective immunohistochemical staining of blood and lymphatic vessels reveals independent prognostic influence of blood and lymphatic vessel invasion in early-stage cervical cancer. Clin Cancer Res 7:93–97

Bunone G, Vigneri P, Mariani L et al 1999 Expression of angiogenesis stimulators and inhibitors in human thyroid tumors and correlation with clinical pathological features. Am J Pathol 155:1967–1976

Byzova TV, Goldman CK, Jankau J et al 2002 Adenovirus encoding vascular endothelial growth factor-D induces tissue-specific vascular patterns in vivo. Blood 99:4434–4442

de Waal RM, van Altena MC, Erhard H, Weidle UH, Nooijen PT, Ruiter DJ 1997 Lack of lymphangiogenesis in human primary cutaneous melanoma. Consequences for the mechanism of lymphatic dissemination. Am J Pathol 150:1951–1957

Dumont DJ, Jussila L, Taipale J et al 1998 Cardiovascular failure in mouse embryos deficient in VEGF receptor-3. Science 282:946–949

Enholm B, Karpanen T, Jeltsch M et al 2001 Adenoviral expression of vascular endothelial growth factor-C induces lymphangiogenesis in the skin. Circ Res 88:623–629

Fielder W, Graeven U, Ergun S et al 1997 Expression of FLT4 and its ligand VEGF-C in acute myeloid leukemia. Leukemia 11:1234–1237

Furudoi A, Tanaka S, Haruma K et al 2002 Clinical significance of vascular endothelial growth factor C expression and angiogenesis at the deepest invasive site of advanced colorectal carcinoma. Oncology 62:157–166

Gale NW, Thurston G, Hackett SF et al 2002 Angiopoietin-2 is required for postnatal angiogenesis and lymphatic patterning, and only the latter role is rescued by angiopoietin-1. Dev Cell 3:411–423

Gannon G, Mandriota SJ, Cui L, Baetens D, Pepper MS, Christofori G 2002 Overexpression of vascular endothelial growth factor-A165 enhances tumor angiogenesis but not metastasis during beta-cell carcinogenesis. Cancer Res 62:603–608

George ML, Tutton MG, Janssen F et al 2001 VEGF-A, VEGF-C, and VEGF-D in colorectal cancer progression. Neoplasia 3:420–427

Gunningham SP, Currie MJ, Han C et al 2000 The short form of the alternatively spliced flt-4 but not its ligand vascular endothelial growth factor C is related to lymph node metastasis in human breast cancers. Clin Cancer Res 6:4278–4286

Hanahan D 1985 Heritable formation of pancreatic beta-cell tumours in transgenic mice expressing recombinant insulin/simian virus 40 oncogenes. Nature 315:115–122

Hartveit E 1990 Attenuated cells in breast stroma: the missing lymphatic system of the breast. Histopathology 16:533–543

Hashimoto I, Kodama J, Seki N et al 2001 Vascular endothelial growth factor-C expression and its relationship to pelvic lymph node status in invasive cervical cancer. Br J Cancer 85:93–97

He Y, Kozaki K, Karpanen T et al 2002 Suppression of tumor lymphangiogenesis and lymph node metastasis by blocking vascular endothelial growth factor receptor 3 signaling. J Natl Cancer Inst 94:819–825

Hirai M, Nakagawara A, Oosaki T, Hayashi Y, Hirono M, Yoshihara T 2001 Expression of vascular endothelial growth factors (VEGF-A/VEGF-1 and VEGF-C/VEGF-2) in postmenopausal uterine endometrial carcinoma. Gynecol Oncol 80:181–188

Ichikura T, Tomimatsu S, Ohkura E, Mochizuki H 2001 Prognostic significance of the expression of vascular endothelial growth factor (VEGF) and VEGF-C in gastric carcinoma. J Surg Oncol 78:132–137

Jeltsch M, Kaipainen A, Joukov V et al 1997 Hyperplasia of lymphatic vessels in VEGF-C transgenic mice. Science 276:1423–1425

Kabashima A, Maehara Y, Kakeji Y, Sugimachi K 2001 Overexpression of vascular endothelial growth factor C is related to lymphogenous metastasis in early gastric carcinoma. Oncology 60:146–150

Kajita T, Ohta Y, Kimura K et al 2001 The expression of vascular endothelial growth factor C and its receptors in non-small cell lung cancer. Br J Cancer 85:255–260

Karkkainen M J, Saaristo A, Jussila L et al 2001 A model for gene therapy of human hereditary lymphedema. Proc Natl Acad Sci USA 98:12677–12682

Karpanen T, Egeblad M, Karkkainen M J et al 2001 Vascular endothelial growth factor C promotes tumor lymphangiogenesis and intralymphatic tumor growth. Cancer Res 61:1786–1790

Kinoshita J, Kitamura K, Kabashima A, Saeki H, Tanaka S, Sugimachi K 2001 Clinical significance of vascular endothelial growth factor-C (VEGF-C) in breast cancer. Breast Cancer Res Treat 66:159–164

Kirkin V, Mazitschek R, Krishnan J et al 2001 Characterization of indolinones which preferentially inhibit VEGF-C- and VEGF-D-induced activation of VEGFR-3 rather than VEGFR-2. Eur J Biochem 268:5530–5540

Kitadai Y, Amioka T, Haruma K et al 2001 Clinicopathological significance of vascular endothelial growth factor (VEGF)-C in human esophageal squamous cell carcinomas. Int J Cancer 93:662–666

Komuro H, Kaneko S, Kaneko M, Nakanishi Y 2001 Expression of angiogenic factors and tumor progression in human neuroblastoma. J Cancer Res Clin Oncol 127:739–743

Kriehuber E, Breiteneder-Geleff S, Groeger M et al 2001 Isolation and characterization of dermal lymphatic and blood endothelial cells reveal stable and functionally specialized cell lineages. J Exp Med 194:797–808

Kubo H, Cao R, Brakenhielm E, Makinen T, Cao Y, Alitalo K 2002 Blockade of vascular endothelial growth factor receptor-3 signaling inhibits fibroblast growth factor-2-induced lymphangiogenesis in mouse cornea. Proc Natl Acad Sci USA 99:8868–8873

Kubo H, Fujiwara T, Jussila L et al 2000 Involvement of vascular endothelial growth factor receptor-3 in maintenance of integrity of endothelial cell lining during tumor angiogenesis. Blood 96:546–553

Kurebayashi J, Otsuki T, Kunisue H et al 1999 Expression of vascular endothelial growth factor (VEGF) family members in breast cancer. Jpn J Cancer Res 90:977–981

Leu A J, Berk DA, Lymboussaki A, Alitalo K, Jain RK 2000 Absence of functional lymphatics within a murine sarcoma: a molecular and functional evaluation. Cancer Res 60:4324–4327

Makinen T, Jussila L, Veikkola T et al 2001a Inhibition of lymphangiogenesis with resulting lymphedema in transgenic mice expressing soluble VEGF receptor-3. Nat Med 7:199–205

Makinen T, Veikkola T, Mustjoki S et al 2001b Isolated lymphatic endothelial cells transduce growth, survival and migratory signals via the VEGF-C/D receptor VEGFR-3. EMBO J 20:4762–4773

Mandriota S J, Jussila L, Jeltsch M et al 2001 Vascular endothelial growth factor-C-mediated lymphangiogenesis promotes tumour metastasis. EMBO J 20:672–682

Mattila MM, Ruohola JK, Karpanen T, Jackson DG, Alitalo K, Harkonen PL 2002 VEGF-C induced lymphangiogenesis is associated with lymph node metastasis in orthotopic MCF-7 tumors. Int J Cancer 98:946–951

Narko K, Enholm B, Makinen T, Ristimaki A 1999 Effect of inflammatory cytokines on the expression of the vascular endothelial growth factor-C. Int J Exp Pathol 80:109–112

Nathanson SD, Zarbo RJ, Wachna DL, Spence CA, Andrzejewski TA, Abrams J 2000 Microvessels that predict axillary lymph node metastases in patients with breast cancer. Arch Surg 135:586–594

Niki T, Iba S, Tokunou M, Yamada T, Matsuno Y, Hirohashi S 2000 Expression of vascular endothelial growth factors A, B, C, and D and their relationships to lymph node status in lung adenocarcinoma. Clin Cancer Res 6:2431–2439

Niki T, Iba S, Yamada T, Matsuno Y, Enholm B, Hirohashi S 2001 Expression of vascular endothelial growth factor receptor 3 in blood and lymphatic vessels of lung adenocarcinoma. J Pathol 193:450–457

O-Charoenrat P, Rhys-Evans P, Eccles SA 2001 Expression of vascular endothelial growth factor family members in head and neck squamous cell carcinoma correlates with lymph node metastasis. Cancer 92:556–568

Oh SJ, Jeltsch MM, Birkenhager R et al 1997 VEGF and VEGF-C: specific induction of angiogenesis and lymphangiogenesis in the differentiated avian chorioallantoic membrane. Dev Biol 188:96–109

Ohta Y, Shridhar V, Bright RK et al 1999 VEGF and VEGF type C play an important role in angiogenesis and lymphangiogenesis in human malignant mesothelioma tumours. Br J Cancer 81:54–61

Ohta Y, Nozawa H, Tanaka Y, Oda M, Watanabe Y 2000 Increased vascular endothelial growth factor and vascular endothelial growth factor-c and decreased nm23 expression associated with microdissemination in the lymph nodes in stage I non-small cell lung cancer. J Thorac Cardiovasc Surg 119:804–813

Okamoto M, Nishimine M, Kishi M et al 2002 Prediction of delayed neck metastasis in patients with stage I/II squamous cell carcinoma of the tongue. J Oral Pathol Med 31:227–233

Orpana A, Salven P 2002 Angiogenic and lymphangiogenic molecules in hematological malignancies. Leuk Lymphoma 43:219–224

Paavonen K, Puolakkainen P, Jussila L, Jahkola T, Alitalo K 2000 Vascular endothelial growth factor receptor-3 in lymphangiogenesis in wound healing. Am J Pathol 156:1499–1504

Padera TP, Kadambi A, di Tomaso E et al 2002 Lymphatic metastasis in the absence of functional intratumor lymphatics. Science 296:1883–1886

Papoutsi M, Siemeister G, Weindel K et al 2000 Active interaction of human A375 melanoma cells with the lymphatics in vivo. Histochem Cell Biol 114:373–385

Papoutsi M, Sleeman JP, Wilting J 2001 Interaction of rat tumor cells with blood vessels and lymphatics of the avian chorioallantoic membrane. Microsc Res Tech 55:100–107

Pepper MS 2001 Lymphangiogenesis and tumor metastasis: myth or reality? Clin Cancer Res 7:462–468

Petrova TV, Makinen T, Makela TP et al 2002 Lymphatic endothelial reprogramming of vascular endothelial cells by the Prox-1 homeobox transcription factor. EMBO J 21:4593–4599

Podgrabinska S, Braun P, Velasco P et al 2002 Molecular characterization of lymphatic endothelial cells. Proc Natl Acad Sci USA 99:16069–16074

Saaristo A, Veikkola T, Tammela T et al 2002 Lymphangiogenic gene therapy with minimal blood vascular side effects. J Exp Med 196:719–730

Salven P, Lymboussaki A, Heikkila P et al 1998 Vascular endothelial growth factors VEGF-B and VEGF-C are expressed in human tumors. Am J Pathol 153:103–108

Schoppmann SF, Birner P, Studer P, Breiteneder-Geleff S 2001a Lymphatic microvessel density and lymphovascular invasion assessed by anti-podoplanin immunostaining in human breast cancer. Anticancer Res 21:2351–2355

Schoppmann SF, Schindl M, Breiteneder-Geleff S et al 2001b Inflammatory stromal reaction correlates with lymphatic microvessel density in early-stage cervical cancer. Anticancer Res 21:3419–3423

Schoppmann SF, Birner P, Stockl J et al 2002 Tumor-associated macrophages express lymphatic endothelial growth factors and are related to peritumoral lymphangiogenesis. Am J Pathol 161:947–956

Skobe M, Hawighorst T, Jackson DG et al 2001a Induction of tumor lymphangiogenesis by VEGF-C promotes breast cancer metastasis. Nat Med 7:192–198

Skobe M, Hamberg LM, Hawighorst T et al 2001b Concurrent induction of lymphangiogenesis, angiogenesis, and macrophage recruitment by vascular endothelial growth factor-C in melanoma. Am J Pathol 159:893–903

Sleeman JP, Krishnan J, Kirkin V, Baumann P 2001 Markers for the lymphatic endothelium: in search of the holy grail? Microsc Res Tech 55:61–69

Stacker SA, Caesar C, Baldwin ME et al 2001 VEGF-D promotes the metastatic spread of tumor cells via the lymphatics. Nat Med 7:186–191

Stacker SA, Achen MG, Jussila L, Baldwin ME, Alitalo K 2002 Metastasis: lymphangiogenesis and cancer metastasis. Nat Rev Cancer 2:573–583

Takahashi A, Kono K, Itakura J et al 2002 Correlation of vascular endothelial growth factor-C expression with tumor-infiltrating dendritic cells in gastric cancer. Oncology 62:121–127

Tanaka K, Sonoo H, Kurebayashi J et al 2002a Inhibition of infiltration and angiogenesis by thrombospondin-1 in papillary thyroid carcinoma. Clin Cancer Res 8:1125–1131

Tanaka K, Kurebayashi J, Sonoo H et al 2002b Expression of vascular endothelial growth factor family messenger RNA in diseased thyroid tissues. Surg Today 32:761–768

Tang RF, Itakura J, Aikawa T et al 2001 Overexpression of lymphangiogenic growth factor VEGF-C in human pancreatic cancer. Pancreas 22:285–292

Tsurusaki T, Kanda S, Sakai H et al 1999 Vascular endothelial growth factor-C expression in human prostatic carcinoma and its relationship to lymph node metastasis. Br J Cancer 80:309–313

Ueda M, Terai Y, Yamashita Y et al 2002 Correlation between vascular endothelial growth factor-C expression and invasion phenotype in cervical carcinomas. Int J Cancer 98:335–343

Veikkola T, Jussila L, Makinen T et al 2001 Signalling via vascular endothelial growth factor receptor-3 is sufficient for lymphangiogenesis in transgenic mice. EMBO J 20:1223–1231

White JD, Hewett PW, Kosuge D et al 2002 Vascular endothelial growth factor-D expression is an independent prognostic marker for survival in colorectal carcinoma. Cancer Res 62:1669–1675

Wigle JT, Oliver G 1999 Prox1 function is required for the development of the murine lymphatic system. Cell 98:769–778

Yang W, Klos K, Yang Y, Smith TL, Shi D, Yu D 2002 ErbB2 overexpression correlates with increased expression of vascular endothelial growth factors A, C, and D in human breast carcinoma. Cancer 94:2855–2861

Yokoyama Y, Sato S, Futagami M et al 2000 Prognostic significance of vascular endothelial growth factor and its receptors in endometrial carcinoma. Gynecol Oncol 77:413–418

Yonemura Y, Endo Y, Fujita H et al 1999 Role of vascular endothelial growth factor C expression in the development of lymph node metastasis in gastric cancer. Clin Cancer Res 5:1823–1829

Yonemura Y, Fushida S, Bando E et al 2001 Lymphangiogenesis and the vascular endothelial growth factor receptor (VEGFR)-3 in gastric cancer. Eur J Cancer 37:918–923

Yuan L, Moyon D, Pardanaud L et al 2002 Abnormal lymphatic vessel development in neuropilin 2 mutant mice. Development 129:4797–4806

DISCUSSION

Blankenstein: Was the glucose level and survival relationship different between the single and double construct mice?

Pepper: It was the same. It has no effect on the function of the islets.

Harris: I liked the histology showing the insulin was near the growing edge of the tumour. This seemed to be enhanced when lymphatics were near the edge also. Could there be any feedback from the lymphatic endothelium to the tumour cells at that interface?

Pepper: Yes, this is a good working hypothesis, and we would predict that the lymphatic endothelium produces not only chemotactic but also trophic factors. This is very easy to test now that we have pure populations of lymphatic endothelial cells and tumour cells. It can be done in co-culture and also *in vivo* through laser microdissection and the use of arrays.

Harris: You didn't really get the invasion until you had the lymphatics there. Could it be that the lymphatics were encouraging the tumour cells to invade when they weren't before?

Pepper: I would agree completely with that hypothesis.

Harris: The heterogeneity of the insulin expression is also of interest. You have genetic lesions; every cell potentially has the insulin promoter switched on. Yet you only see this rim. How do you explain this?

Pepper: I can't. Clearly, the interaction of the tumour cells with the stromal cells and lymphoid cells seems important. The stromal cells could be producing factors which maintain the differentiated state of these endocrine cells. Exactly what they are and whether this actually occurs is unclear.

Balkwill: I was interested that when you showed the pictures of the tumour cells in the lymph nodes, they were actually in clumps. It looked like there was quite a hefty clump of cells there. Is that normal?

Pepper: The cells very rarely migrate as individual cells in this model. They are usually in aggregates. These might be aggregates that seed out into the subcapsular region and grow locally, or there may be single cells that we are not detecting which then seed out.

Balkwill: So they can move into the lymphatic system and grow there, before moving into the lymph node.

Pepper: One of the things we did see in these mice was that the number of tumours was increased in the double-transgenic mice. There are two interpretations. One is that there is an angiogenic effect that we were unable to detect. The other is that as these tumour cell masses move through the lymphatics, and some of them are lodged within lymphatics in the pancreas, they then grow within the lymphatics. This would give us a false impression that there are more primary tumours, whereas these may have originated from the metastatic tumour cell masses.

Balkwill: How do you think the tumour cells get from the lymphatics just outside the tumour cell area into the lymph node? Is this just a mechanical process?

Pepper: The structure of lymphatic capillaries gives us a clue: these capillaries have no pericytes and no smooth muscle cells. The idea is that it is just an increase in interstitial fluid pressure which forces the cells to move down. As they get further down towards the collecting ducts and lymph nodes, there are contractile cells in the wall of the lymphatics. This could then push the cells further. Initially, it is probably just interstitial fluid pressure and an increase in lymph flow.

Balkwill: Do you think that chemokines would kick in later on?

Pepper: No, I think they come in earlier. They are likely to be involved in the initial invasion into the lymphatics.

Cerundolo: Have you done any experiments to look at the tumour-specific immune response in these mice?

Pepper: We are not set up to do this. People have often asked me whether immune function is altered in these mice, but we haven't looked at this.

Blankenstein: This is probably not possible in the RIP-Tag tumours, because they are described as tolerant. You would probably have to do this experiment in RIP-Tag5 mice, where the Tag antigen is expressed first time in adult mice.

Gordon: Could you tell us about the different RIP-Tag models?

Blankenstein: Hanahan made a couple of RIP-Tag mice that express SV40 as oncogene under the insulin promoter. In the RIP-Tag2 mouse SV40 is expressed early in ontogeny and the mice are tolerant for Tag. In the RIP-Tag5 mice SV40 is expressed from week 6 in age. These mice are, at least partially, immunocompetent against the TAG antigen.

Rollins: Along the lines of Enzo Cerundolo's question, do the insulinomas in the RIP-Tag mice elicit an inflammatory infiltrate? If so, is the nature of the infiltrate altered in the double transgenics?

Pepper: There is no obvious inflammatory infiltrate, either in the RIP-Tag or the double transgenic mice.

Gordon: We have seen some macrophages. It depends on the marker we use, macrosialin, but not F4/80.

Richmond: Which chemokines does VEGF-C induce?

Pepper: This is ongoing work. We have run some arrays, but we have yet to confirm the data we have. There are at least two other groups who are doing the same work.

Richmond: Mary Hendrix has postulated that tumour cells have the ability to form these little vessels which then might merge with the lymphatics. Do you see any evidence with this?

Pepper: We see a lot of haemorrhage in these tumours. The question has always been whether these so called 'vascular channels' are devoid of endothelium. It is very difficult to study this. I think the jury is still out on this question. Most people would say that you cannot have functional vascular channels without endothelium, but we do see lakes of blood in these tumours which don't have endothelium. It is possible.

Oppenheim: We have used VEGF to induce CXCR4. This is the only chemokine receptor that we know of that goes up in response to VEGF. I don't whether it is VEGF-A, B, C or D.

Pepper: It is probably A.

Mantovani: Why is it that lymphatic endothelial cells do not penetrate the tumour or even the normal islets? The pictures were striking.

Pepper: It is not a trivial question. We don't have the answer, but one could think in terms of ECM. There may also be some anti-lymphangiogenic factors being produced by islet cells. The question to me has always been why are there no lymphatics in the brain? The CNS is completely devoid of lymphatics. Yet we have looked in every region of the brain that we can dissect, and we find VEGF-C in all of them.

D'Incalci: There has been some interest in developing approaches to bind some anti-cancer drugs to macromolecules. I heard that one of the rationales for this approach was that the lack of intratumour lymphatics would decrease the clearance from the tumour in a selective way. Do you think this approach is sensible?

Pepper: It is sensible. The only problem is that interstitial fluid pressure in tumours is purported to be very high. The difficulty will be getting the drug from the vascular system into the tumour interstitium in the first place. This is where the limitation comes. Once you can do that, it would stay in the interstitium. But you have to get it there in the first place.

Balkwill: You showed us a nice picture of the lymphatics in inflammation. They look different here. The lymphatics do move into the whole tissue in inflammation. Is it true that in human and mouse tumours you see lymphatics around the edge, but if you have an inflammatory site they go into the tissue?

Pepper: Yes, it happens in wound healing and inflammation. The only point to make here is that lymphangiogenesis is delayed relative to angiogenesis. Angiogenesis gets going first and then 24–48 h later lymphangionesis is induced.

Ristimäki: You showed a long list of clinical work associating VEGF-C with lymphatic spread. Would VEGF-C give any benefit for staging tumours, in comparison, for instance, with finding tumour cells in the lymph nodes per se? Would the expression of VEGF-C be a helpful staging criterion?

Pepper: There is some work going on in some gynaecological tumours showing that there may be a switch. This work is about to be submitted. I don't think there are any published data that have actually looked at progression.

Balkwill: What induces VEGF-C, and is it high in inflammation also?

Pepper: It is very high in inflammation. Unlike classical VEGF, which is induced by hypoxia, VEGF-C is not induced by hypoxia. But it is induced by inflammatory cytokines such as IL1 and TNF.

Van Trappen: With regard to lymphangiogenesis and inflammation, there was a recent paper from Kerjaschki's group (Schoppmann et al 2002) showing that macrophages at the periphery of the tumour are expressing VEGF-C mRNA as well as protein. This suggests that there may be a role for it in the beginning and early invasion in cervical cancer, and inflammatory cells might induce lymphangiogenesis besides the tumour cells.

Blankenstein: You finished with the somewhat pessimistic (and perhaps realistic) notion that you don't believe that the inhibition of tumour-induced lymphangiogenesis is a therapeutic option. Can you explain why you are so pessimistic?

Pepper: The reason for the pessimism is that when patients present with tumours, if one can detect a primary tumour in most situations this is removed. If the metastases are present the metastatic process has already occurred and there is nothing you can do about this, they are there. Removing the primary tumour will remove the source of the cells that can metastasize. At presentation, unless you actually leave the primary tumour in place we are not in a position to prevent metastases. They have already occurred. The key issue here is not to prevent metastases but to maintain them in a dormant state. This is the real therapeutic window that we are looking at.

Van Trappen: Yesterday the question about which patients we are going to treat with anti-inflammatory agents was raised. Many groups have looked extensively at micrometastases in breast cancer. At initial presentation, we know that 30% of breast cancer patients with node-negative disease develop recurrent disease, and subsequently have decreased survival. The hypothesis was raised that these patients might have micrometastases at initial presentation. Clinical studies have shown that 30–40% of patients with early stage breast cancer have indeed micrometastases in the blood and/or bone marrow. Subsequent work is now ongoing investigating why micrometastases stay dormant for a time and then become active. More research into this interesting topic is required to understand recurrent disease. Certain characteristics of these dormant tumour cells that regrow *in vitro* have been identified. Most of these micrometastases are down-regulating MHC class I antigens and are non-proliferative. We found recently that in early stage cervical cancer (70% of cervical cancers) about 40–50% have micrometastases in the lymph nodes. As VEGF-C is crucial in lymphatic spread, blocking its function might have an effect on early spread of micrometastases into the lymphatic system. In cervical and breast cancer at least, metastasis seems to occur very early, therefore, it is important to explore further what drives dormant micrometastases to start growing again. It is suggestive that, in this latter process, the microenvironment plays a crucial role. In ovarian cancer patients, when we give chemotherapy we see a response in 70–80% of patients. In the remaining group (20%), resistance to chemotherapy is due to multidrug resistance genes, but also there is the possibility that lots of these tumour cells have down-regulation of MHC class I antigens (these cells don't respond to chemotherapy). After a variable length of time a subclone of these cells starts to grow again. We should explore further the role of inflammatory cells in activating these dormant tumour cells.

Blankenstein: This work pioneered by G. Riethmüller is certainly great work, but we have to consider that we have access only to certain organs, for example bone marrow, where cytokeratin-positive tumour cells, named micrometastases, can be found. Such cells can serve as prognostic markers for poor prognosis. However, the metastases that become clinically apparent later on occur somewhere else. Whether or not they originate from the bone marrow metastases is an open question. I would also like to comment on the down-regulation MHC class I expression. Usually this is taken as evidence for T-cell involvement and immune selection. We must keep in mind that class I expression is under the control of many factors, for example cell-cycle control. If these cells express lower MHC class I levels, it does not necessarily mean that this is a sign of immune selection.

Gordon: What about the relationship between the endothelium of the vessels and the lymphatics? Don't they sometimes go together? I was under the impression that some of the VEGFs will promote both.

Pepper: Under normal circumstances the two systems don't anastomose.

Gordon: But they seem to coexist.

Pepper: Yes. There are two papers looking at gene expression in the two different cell lines using Affymetrix chips. What is clear is that there are trophic factors and cytokines produced by both cell lines which may be involved in the maintenance of the other cell line. For example, blood endothelial cells make a lot of VEGF-C, which may be required for the maintenance of lymphatics, which are situated in close apposition. There is possible cross-talk between the two. These two systems run in parallel.

Reference

Schoppmann SF, Birner P, Stockl J et al 2002 Tumor-associated macrophages express lymphatic endothelial growth factors and are related to peritumoral lymphangiogenesis. Am J Pathol 161:947–956

Infiltration of tumours by macrophages and dendritic cells: tumour-associated macrophages as a paradigm for polarized M2 mononuclear phagocytes

Alberto Mantovani*†, Silvano Sozzani*‡, Massimo Locati†, Tiziana Schioppa*, Alessandra Saccani*, Paola Allavena* and Antonio Sica*

*Istituto di Ricerche Farmacologiche Mario Negri, Via Eritrea 62, 20157 Milan and †Centro IDET, Institute of General Pathology, University of Milan, Via Mangiagalli 31, 20133 Milan and ‡Department of Biotechnology, Section of General Pathology and Immunology, University of Brescia, 25123 Brescia, Italy

Abstract. Macrophages and dendritic cells infiltrate tumours. In the tumour microenvironment, mononuclear phagocytes acquire properties of polarized M2 (or alternatively activated) macrophages. These functionally polarized cells, and similarly oriented or immature dendritic cells present in tumours, play a key role in subversion of adaptive immunity and in inflammatory circuits which promote tumour growth and progression.

2004 Cancer and inflammation. Wiley, Chichester (Novartis Foundation Symposium 256) p 137–148

Cells belonging to the monocyte–macrophage lineage have long been recognized to be heterogeneous. For instance, macrophages obtained from certain anatomical sites exhibit peculiar morphology and function as exemplified by lung alveolar macrophages. Since lineage-defined subsets have not been identified to date, macrophage heterogeneity is likely to reflect the plasticity and versatility of these cells in response to exposure to microenvironmental signals. Cytokines and microbial products profoundly affect the function of mononuclear phagocytes. In particular, cytokines associated with polarized type II responses (interleukin [IL]4, IL13, IL10) induce an alternative activation program in macrophages. 'Alternatively activated' (Goerdt & Orfanos 1999, Stein et al 1992) or M2 macrophages show distinct functional properties which integrate them in polarized type II responses.

Macrophages are a major component of the leukocyte infiltrate of tumours (Balkwill & Mantovani 2001, Mantovani et al 1992). Tumour-associated macrophages (TAMs) originate from circulating blood monocytes. Their recruitment and survival *in situ* is directed by chemokines (Mantovani et al 1992)

and by cytokines which interact with tyrosine kinase receptors. TAMs have complex dual functions in their interaction with neoplastic cells (the 'macrophage balance' hypothesis, Mantovani et al 1992) but strong evidence suggests that they are part of inflammatory circuits that promote tumour progression (Balkwill & Mantovani 2001, Mantovani et al 1992).

Here we will concisely review the properties of polarized macrophages and summarize recent information consistent with the view that TAMs are a polarized type II (or M2, or alternatively activated) macrophage population (Goerdt & Orfanos 1999, Mantovani et al 2002, Mills et al 2000, Stein et al 1992).

Functional polarization of macrophages

The properties of polarized macrophages have recently been reviewed (Mantovani et al 2002). They include differential cytokine and chemokine production and effector function. In analogy with the Th1/Th2 dichotomy, polarized macrophage populations have been called M1 (typically activated by interferon [IFN]γ + lipopolysaccharide [LPS]) and M2 (or alternatively activated, exposed to IL4 and IL13).

Different inducers of type II polarization, elicit distinct functional phenotypes in macrophages. Given the fact that inducers such as IL4/IL13, IL10, glucocorticoid hormones and vitamin D3 use widely different receptors and signalling pathways, this is hardly surprising. Moreover, mononuclear phagocytes are exposed to a multiplicity of signals *in vivo* with different temporal patterns. Therefore, polarization of macrophage function should be viewed as an operationally useful, simplified conceptual framework describing a continuum of diverse functional states. With this general caveat, available information suggests that classically activated type I macrophages are potent effector cells which kill microorganisms and tumour cells and produce copious amounts of proinflammatory cytokines. In contrast, type II macrophages tune inflammatory responses and adaptive Th1 immunity, scavenge debris, and promote angiogenesis, tissue remodelling and repair.

Recruitment of TAMs

TAMs derive from circulating monocytic precursors and *in situ* proliferation is generally not an important mechanism that sustains the mononuclear phagocyte population, at least in human tumours (Mantovani et al 1992). Several lines of evidence, including correlation between production and infiltration in murine and human tumours, passive immunization and gene modification, indicate that chemokines play a pivotal role in the recruitment of monocytes in neoplastic tissues (Rollins 1999). Chemokines are usually classified according to their constitutive (e.g. CXCL12) or inducible production (e.g. CCL2, CXCL8).

Tumours are generally characterized by the constitutive expression of chemokines belonging to the inducible realm (Mantovani 1999). The molecular mechanisms accounting for constitutive expression have been defined only for CXCL1 and involve NF-κB activation. Melanoma cells display high expression of NF-κB-inducing kinase (NIK) (Yang & Richmond 2001), constitutive activation of IκB kinase activity and MAPK signalling cascades, as well as constitutive activation of NF-κB (Dhawan & Richmond 2002). This may represent a general mechanism underlying constitutive expression of inflammatory chemokines in tumours.

A variety of chemokines (CCL2, CCL5, CXCL12, CXCL8, CXCL1, CXCL13, CCL17 and CCL22) have been detected as products of cancer cells or tumour stromal elements. For review, see Mantovani et al (2003). CXCL1 and related molecules (CXCL2, CXCL3, CXCL8 or IL8) have been shown to play an important role in melanoma progression (Haghnegahdar et al 2000) by stimulating neoplastic growth, promotion of inflammation and induction of angiogenesis.

Vascular endothelial growth factor (VEGF) and macrophage colony stimulating factor (M-CSF) are cytokines commonly produced by tumours, which interact with tyrosine kinase receptors and elicit monocyte migration. These molecules can contribute to macrophage recruitment in tumours (Barleon et al 1996, Duyndam et al 2002, Lin et al 2001). Studies in M-CSF-deficient mice (*op/op*) have strongly supported the concept of the pro-tumour function of the mononuclear phagocyte system. M-CSF deficiency in *op/op* mice diminishes macrophage recruitment, stroma formation and tumour growth in the Lewis lung carcinoma model (Nowicki et al 1996). In a mammary carcinoma model, M-CSF deficiency did not affect early stages of tumour development, but reduced progression to invasive carcinoma and metastasis (Lin et al 2001). Genetic restoration of M-CSF production in epithelial cells restored macrophage infiltration as well as malignant behaviour.

Intratumour survival and differentiation of TAMs

Macrophage survival in tumours may contribute to the levels of infiltration. CSFs, and M-CSF in particular, are likely to promote the macrophage lifespan as well as, in some murine tumours, proliferation of TAMs (Barleon et al 1996, Duyndam et al 2002, Lin et al 2001, Mantovani et al 1992) (Fig. 1). Recently, placenta-derived growth factor (PlGF), a molecule related to VEGF in terms of structure and receptor usage, has been reported to promote the survival of TAMs (Adini et al 2002).

TAMs isolated from murine tumours and from human ovarian cancer express very low levels of inflammatory chemokine receptors, CCR2 in particular (Sica et al 2000a). Primary inflammatory signals, including tumour necrosis factor (TNF)α,

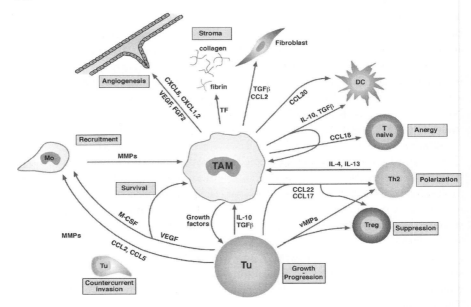

FIG. 1.   TAMs as polarized type II macrophages. TAMs are a source and target for cytokines and chemokines in the tumour microenvironment. These molecules and other local mediators (e.g. tissue factor, TF; MMPs) regulate tumour growth, progression and invasion, interaction with other components of the stroma, and the activation and orientation of adaptive immunity (modified from Mantovani et al 2002).

which is expressed in the ovarian carcinoma microenvironment (Naylor et al 1993), down-regulate CCR2 in monocytes (Sica et al 1997). Therefore, down-regulation of CCR2 may in part reflect the monocyte-to-macrophage transition and, in part, may serve as a mechanism to localize and retain macrophages in tumours.

Cytokines have the potential to promote and orient the differentiation of recruited mononuclear phagocytes. IL10 as well as transforming growth factor (TGF)β are produced by a variety of tumour cells (including ovarian cancer) and by TAMs themselves (Kambayashi et al 1995, Kim et al 1995, Loercher et al 1999, Maeda et al 1995, Sica et al 2000b). IL10 promotes the differentiation of monocytes to mature macrophages and blocks their differentiation to dendritic cells (DCs) (Allavena et al 2000). Thus a gradient of IL10 into the tumour tissue may be an important determinant of the relative proportion of TAMs versus tumour-associated DCs (TADC) and of their relative distribution. For instance, in papillary carcinoma of the thyroid, TAMs are evenly distributed throughout the tissue, in contrast to DCs which are present in the periphery (Scarpino et al 2000). To the extent that they have been investigated, TAMs have phenotype and function similar to type II macrophages. TAMs from poorly immunogenic malignant tumours display decreased cytotoxic activity and they actually promote tumour cell proliferation in particular under suboptimal culture conditions (Mantovani et al 1992). They are poor producers of NO (DiNapoli et al 1996) and, *in situ* in ovarian cancer, a minority of macrophages mainly localized at the periphery scored positive for iNOS (Klimp et al 2001). Arginase expression in TAMs has not been studied. TAMs have high mannose receptor (P. Allavena, unpublished) and are poor at presenting antigen (Mantovani et al 1992). As discussed below, they have an IL10^{high} IL12^{low} cytokine repertoire.

T cells infiltrating various types of human tumours have a type II phenotype, with a predominance of CD8$^+$ (e.g. Kaposi's sarcoma) or CD4$^+$ cells (e.g. cervical carcinoma) in different neoplasms (Balkwill & Mantovani 2001). By producing IL4, IL13 and IL10, tumour-infiltrating T cells may reinforce the skewing of monocyte differentiation in tumours towards a type II phenotype.

Chemokines in the orientation of adaptive immunity

Chemokines are also part of amplification and regulation systems of polarized T cell responses and may activate type II responses or suppression circuits in tumours (Mantovani et al 2002). For instance, the oncogenic virus human herpesvirus 8 (HHV8), involved in the pathogenesis of Kaposi's sarcoma and haematological malignancies, encodes three CC chemokines (vMIPI, II and III) which interact as agonists with CCR3, CCR4 and CCR8 and, accordingly, preferentially attract polarized type II T cells (Balkwill & Mantovani 2001) and, presumably, T$_{reg}$ cells (Iellem et al 2001). In agreement, Kaposi's sarcoma is infiltrated by CD8$^+$ and, to a lesser extent, CD4$^+$ cells with a predominant type II phenotype. Therefore HHV8 virus-encoded chemokines represent a strategy to subvert effective antiviral/antitumour immunity by favouring type II responses and, possibly, T$_{reg}$ cells.

TAMs express a selected set of these mediators (CCL2, CCL22, CCL18). In particular, tumour cell- and macrophage-derived IL10 likely accounts for CCL18 production by TAMs in human ovarian ascites fluid (Schutyser et al 2002). By interacting with an unidentified receptor (Goerdt & Orfanos 1999), CCL18 attracts naïve T cells in a peripheral microenvironment dominated by type II macrophages and immature DCs and may induce anergy.

IL10 and TGFβ

TAMs have poor antigen presenting capacity and can suppress T cell activation and proliferation (Balkwill & Mantovani 2001). The suppressive mediators produced by TAM include prostaglandins, IL10 and TGFβ (Kambayashi et al 1995, Kim et al 1995, Loercher et al 1999, Maeda et al 1995, Sica et al 2000b). Moreover, they do not produce IL12 spontaneously and are refractory to stimulation by IFNγ and LPS. The IL10^{high} IL12^{low} phenotype is characteristic of polarized type II macrophages (see above).

IL10 derived from tumour cells and T cells may favour differentiation along this pathway, as discussed above. Antibody blocking experiments suggest that autocrine IL10 in part accounts for the defective IL12 production (Sica et al 2000b). Ibe et al (2001) have recently suggested that during tumour establishment T cells condition TAM to produce IL10 and that inactivation of T cells results in a switch of TAMs towards IFNγ production and elicits tumour rejection.

Analysis of the molecular basis responsible for defective IL12 production in TAMs (Sica et al 2000b) revealed that they display a massive and constitutive nuclear localization of the NF-κB inhibitory p50 homodimer (A. Sica, unpublished), which most likely provides a molecular mechanism for other alterations of TAM functions dependent on NF-κB activation, including defective iNOS expression and NO production (DiNapoli et al 1996, Klimp et al 2001). Further clarification of the L-arginine metabolism in TAMs may be of particular relevance, as it was recently observed that the activity of this metabolic pathway in suppressor myeloid cells controls T cell functions (Bronte et al 2003).

Tumour-associated DCs (TADCs)

The actual significance of TADCs for human tumour progression is uncertain. TADCs generally show an immature phenotype with high CD1a and low co-stimulatory molecules CD80, CD86 and CD40 (Allavena et al 2000, Chaux et al 1996, Mantovani et al 2002, Scarpino et al 2000). Immature DCs would induce tolerance and the generation of T$_{reg}$ cells that will inhibit the generation of cytotoxic T lymphocytes. TADC phenotype may reflect lack of effective maturation signals *in situ*, prompt migration to lymph nodes of mature cells or the presence of maturation inhibitors in the tumour context. Many tumours produce IL10, IL6, TGFβ and M-CSF. IL10 has been shown to block the differentiation and maturation of DCs, including TADCs (Allavena et al 2000). Inhibition of DC differentiation from bone marrow precursors has also been observed with IL6 and M-CSF. One could speculate that TADCs may maintain tolerance to tumour antigens and, in analogy with TAMs (Mantovani et al 1992), may in some tumours promote tumour progression and dissemination. Production of PGE2 and release of ATP by dying cells in the tumour microenvironment might also contribute to the inhibition of DC maturation and IL12 production.

Tissue remodelling, fibrosis and angiogenesis

Phagocytes play a central role in tissue remodelling and repair during ontogenesis and adult life. This ancestral function is expressed by TAMs which produce a host of growth factors affecting tumour cell proliferation, angiogenesis, and the deposition and dissolution of connective tissues. These include epidermal growth factor (EGF), members of the fibroblast growth factor (FGF) family, TGFβ, VEGF and chemokines (Mantovani et al 2002).

Macrophages can produce enzymes and inhibitors which regulate the digestion of the extracellular matrix, such as matrix metalloproteinases (MMPs), plasmin, urokinase-type plasminogen activator (uPA) and the uPA receptor. Interestingly, MMP9 derived from haematopoietic cells of host origin contributes to skin carcinogenesis (Bergers et al 2000, Coussens et al 2000). Chemokines can induce a restricted and distinct transcriptional profiling program in human monocytes (Locati et al 2002), including gene expression of various MMPs and, in particular, MMP9 production, along with the uPA receptor and the cytochrome CYP1B1 involved in carcinogenesis. Induction of these molecules is part of a program of activation which arms monocytes with tools (receptors and enzymes) important for migration in tissues. Evidence suggests that MMP9 has complex effects beyond matrix degradation including promotion of the angiogenesis switch and release of growth factors (Bergers et al 2000, Coussens et al 2000).

Concluding remarks

The view of TAMs as a skewed type II (M2) macrophage population is an oversimplification. Indeed tumours are a diverse set of disorders, and a systematic effort of *in vitro* and *in vivo* characterization has been made only in selected systems (e.g. ovarian and breast cancer for human tumours). Moreover, available information indicates that the distribution and function of TAMs and TADCs differ considerably in different microregions of the neoplastic tissue (peripheral versus central; hypoxic versus normoxic areas). These limitations, as well as technological developments, call for gene expression profiling of infiltrating cells directly obtained from different regions of different tumours. Moreover, little dynamic information, analysing function in relation to stages of progression, is available. In spite of these limitations, the view of TAMs as a polarized type II (M2) macrophage is efficacious in summarizing current understanding of the immunobiology of these cells.

By expressing properties of polarized type II macrophages, TAMs participate in circuits that regulate tumour growth and progression, adaptive immunity, stroma formation and angiogenesis. In particular, TAMs are a key component of inflammatory circuits which promote tumour progression and metastasis (for review see Coussens & Werb 2001). The general hypothesis of a pro-tumour role of inflammation, and of macrophages in particular, is supported by several lines of evidence including genetic analysis and gene targeting. This raises the possibility that the molecules and cells involved may represent novel, valuable therapeutic targets.

References

Adini A, Kornaga T, Firoozbakht F, Benjamin LE 2002 Placental growth factor is a survival factor for tumor endothelial cells and macrophages. Cancer Res 62: 2749–2752

Allavena P, Sica A, Vecchi A, Locati M, Sozzani S, Mantovani A 2000 The chemokine receptor switch paradigm and dendritic cell migration: its significance in tumor tissues. Immunol Rev 177:141–149

Balkwill F, Mantovani A 2001 Inflammation and cancer: back to Virchow? Lancet 357:539–545

Barleon B, Sozzani S, Zhou D, Weich HA, Mantovani A, Marmè D 1996 Migration of human monocytes in response to vascular endothelial growth factor (VEGF) is mediated via the flt-1/VEGF receptor. Blood 87:3336–3343

Bergers G, Brekken R, McMahon G et al 2000 Matrix metalloproteinase-9 triggers the angiogenic switch during carcinogenesis. Nat Cell Biol 2:737–744

Bronte V, Serafini P, Mazzoni A, Segal DM, Zanovello P 2003 L-arginine metabolism in myeloid cells controls T-lymphocyte functions. Trends Immunol 24:302–306

Chaux P, Moutet M, Faivre J, Martin F, Martin M 1996 Inflammatory cells infiltrating human colorectal carcinomas express HLA class II but not B7-1 and B7-2 costimulatory molecules of the T-cell activation. Lab Invest 74:975–983

Coussens LM, Werb Z 2001 Inflammatory cells and cancer: think different! J Exp Med 193:F23–F26

Coussens LM, Tinkle CL, Hanahan D, Werb Z 2000 MMP-9 supplied by bone marrow-derived cells contributes to skin carcinogenesis. Cell 103:481–490

Dhawan P, Richmond A 2002 A novel NF-kappa B-inducing kinase-MAPK signaling pathway up-regulates NF-kappa B activity in melanoma cells. J Biol Chem 277: 7920–7928

DiNapoli MR, Calderon CL, Lopez DM 1996 The altered tumoricidal capacity of macrophages isolated from tumor-bearing mice is related to reduced expression of the inducible nitric oxide synthase gene. J Exp Med 183:1323–1329

Duyndam MC, Hilhorst MC, Schluper HM et al 2002 Vascular endothelial growth factor-165 overexpression stimulates angiogenesis and induces cyst formation and macrophage infiltration in human ovarian cancer xenografts. Am J Pathol 160:537–548

Goerdt S, Orfanos CE 1999 Other functions, other genes: alternative activation of antigen-presenting cells. Immunity 10:137–142

Haghnegahdar H, Du J, Wang D et al 2000 The tumorigenic and angiogenic effects of MGSA/GRO proteins in melanoma. J Leukoc Biol 67:53–62

Ibe S, Qin Z, Schuler T, Preiss S, Blankenstein T 2001 Tumor rejection by disturbing tumor stroma cell interactions. J Exp Med 194:1549–1559

Iellem A, Mariani M, Lang R et al 2001 Unique chemotactic response profile and specific expression of chemokine receptors CCR4 and CCR8 by CD4$^+$CD25$^+$ regulatory T cells. J Exp Med 194:847–853

Kambayashi T, Alexander HR, Fong M, Strassmann G 1995 Potential involvement of IL-10 in suppressing tumor-associated macrophages. Colon-26-derived prostaglandin E2 inhibits TNF-alpha release via a mechanism involving IL-10. J Immunol 154: 3383–3390

Kim J, Modlin RL, Moy RL et al 1995 IL-10 production in cutaneous basal and squamous cell carcinomas. A mechanism for evading the local T cell immune response. J Immunol 155:2240–2247

Klimp AH, Hollema H, Kempinga C, van der Zee AG, de Vries EG, Daemen T 2001 Expression of cyclooxygenase-2 and inducible nitric oxide synthase in human ovarian tumors and tumor-associated macrophages. Cancer Res 61:7305–7309

Lin EY, Nguyen AV, Russell RG, Pollard JW 2001 Colony-stimulating factor 1 promotes progression of mammary tumors to malignancy. J Exp Med 193:727–740

Locati M, Deuschle U, Massardi ML et al 2002 Analysis of the gene expression profile activated by the CC chemokine ligand 5/RANTES and by lipopolysaccharide in human monocytes. J Immunol 168:3557–3562

Loercher AE, Nash MA, Kavanagh JJ, Platsoucas CD, Freedman RS 1999 Identification of an IL-10-producing HLA-DR-negative monocyte subset in the malignant ascites of patients with ovarian carcinoma that inhibits cytokine protein expression and proliferation of autologous T cells. J Immunol 163:6251–6260

Maeda H, Kuwahara H, Ichimura Y, Ohtsuki M, Kurakata S, Shiraishi A 1995 TGF-beta enhances macrophage ability to produce IL-10 in normal and tumor-bearing mice. J Immunol 155:4926–4932

Mantovani A 1999 The chemokine system: redundancy for robust outputs. Immunol Today 20:254–257

Mantovani A, Bottazzi B, Colotta F, Sozzani S, Ruco L 1992 The origin and function of tumor-associated macrophages. Immunol Today 13:265–270

Mantovani A, Sozzani P, Locati D, Allavena P, Sica A 2002 Macrophage polarization: tumor-associated macrophages as a paradigm for polarized M2 mononuclear phagocytes. Trends Immunol, in press

Mills CD, Kincaid K, Alt JM, Heilman MJ, Hill AM 2000 M-1/M-2 macrophages and the Th1/Th2 paradigm. J Immunol 164:6166–6173

Naylor MS, Stamp GW, Foulkes WD, Eccles D, Balkwill FR 1993 Tumor necrosis factor and its receptors in human ovarian cancer. Potential role in disease progression. J Clin Invest 91:2194–2206

Nowicki A, Szenajch J, Ostrowska G et al 1996 Impaired tumor growth in colony-stimulating factor 1 (CSF-1)-deficient, macrophage-deficient op/op mouse: evidence for a role of CSF-1- dependent macrophages in formation of tumor stroma. Int J Cancer 65: 112–119

Rollins B 1999 Chemokines and cancer, Humana Press, Totowa, NJ

Scarpino S, Stoppacciaro A, Ballerini F et al 2000 Papillary carcinoma of the thyroid: hepatocyte growth factor (HGF) stimulates tumor cell to release chemokines active in recruiting dendritic cells. Am J Pathol 156:831–837

Schutyser E, Struyf S, Proost P et al 2002 Identification of biologically active chemokine isoforms from ascitic fluid and elevated levels of CCL18/pulmonary and activation-regulated chemokine in ovarian carcinoma. J Biol Chem 277:24584–24593

Sica A, Saccani A, Borsatti A et al 1997 Bacterial lipopolysaccharide rapidly inhibits expression of C-C chemokine receptors in human monocytes . J Exp Med 185:969–974

Sica A, Saccani A, Bottazzi B et al 2000a Defective expression of the monocyte chemotactic protein-1 receptor CCR2 in macrophages associated with human ovarian carcinoma. J Immunol 164:733–738

Sica A, Saccani A, Bottazzi B et al 2000b Autocrine production of IL-10 mediates defective IL-12 production and NF-kappaB activation in tumor-associated macrophages. J Immunol 164:762–767

Stein M, Keshav S, Harris N, Gordon S 1992 Interleukin 4 potently enhances murine macrophage mannose receptor activity: a marker of alternative immunologic macrophage activation. J Exp Med 176:287–292

Yang J, Richmond A 2001 Constitutive IkappaB kinase activity correlates with nuclear factor-kappaB activation in human melanoma cells. Cancer Res 61:4901–4909

DISCUSSION

Brennan: Is the type II phenotype displayed by TAMs reversible? Could you shift it back to being what you would describe as a type I phenotype that was able to make more pro-inflammatory cytokines and wouldn't induce this anergic state in infiltrating lymphocytes?

Mantovani: Concerning the polarization of macrophages, we have never been able *in vitro* to fully activate them and reverse the phenotype.

Gordon: I don't think it is possible to redifferentiate them.

Brennan: What was the association between this and the second part of your talk? Were you going to bring back the pentraxins?

Mantovani: The connection is a loose one. Here we have a situation in which we hope to use innate immunity to protect cancer patients. There is actually a second connection which I didn't mention, but which might be more direct. It is a little early to talk about these results.

Cerundolo: I have a question about the effect of CCL18 on naïve T cells. You told us that CCL18 anergized naïve T cells. Can you remind me about the evidence that it has no effect on activated and tolerized T cells?

Mantovani: It does not anergize T cells. It recruits naïve T cells. To the limited extent that we have tested this we have confirmed it. The speculation is that if you attract a naïve cell in a non-professional environment, the dogma would be that this would result in anergy. This would be our interpretation for these huge amounts of CCL18.

Cerundolo: With regard to PTX3, what is the mechanism to account for its specificity for *Aspergillus* and *Pseudomonas*?

Mantovani: We have some data suggesting that the major sugar moiety of the conidia competes for binding, but these are preliminary. We have had problems doing real binding experiments. Now for the first time we have a microbial moiety in which we can do plasmonic resonance. This is recent work.

Pollard: I wanted to pick up the idea of 'macrophage education'. Guido Forni, in your work you were treating mice with IL12 which had a negative effect on tumours. Do you think this reverses the local phenotype on macrophages, or is there now a recruitment of a whole new subset of monocytes which then differentiate in a different way under a fresh cytokine environment.

Forni: I have no data on this.

Pollard: Alberto Mantovani, to follow this idea of local education, you have made a case for IL10. It seems to me that there are probably other signals. Matrix is likely to provide a very important signal in the environment of the tumour or a developing organ, and this is also signalling to macrophages to do particular things. Have you any data on this?

Mantovani: We have no data. We focused on IL10 because we had the data with IL10. With anti-IL10 *in vitro* we could get some IL12 production. We tried experiments *in vivo* combining anti-IL10 with IL2 or IL12. We didn't really get very impressive results. Christophe Caux has very interesting results using this strategy. Using the right agonist is important. Concerning the matrix, one of the points that Siamon Gordon made in his introduction concerns recognition of the matrix. Are there pattern recognition receptors that are involved in interaction with the matrix and with the tumour cells themselves? I don't know whether these are driving the polarization of the cells towards a type II phenotype.

Pollard: This is an interesting question, and one that we have been thinking about a bit. We know that the tumour stroma — not so much the cells themselves, but the matrix — can actually change the phenotypes of tumour cells, and potentiate them. We wonder therefore whether those same matrix changes are changing the phenotypes of the associated cells which also then help further drive the tumour's survival. I don't know of any data on that sort of idea.

Gordon: We are beginning to look at proteoglycans. The F4/80 family members are present, but what they do to the activation is currently unknown.

Pollard: We should start to think about the matrix composition, not just in terms of cytokines.

Gordon: Does anyone here know about matrix in tumours? Apart from the breaking down of matrix by the tumour, we have not heard much about it. If we look in the literature we find that matrix is everywhere and very undifferentiated. Most people assume it is just the same, but this is not the case.

Pollard: There are classic studies by Mina Bissel and others who can change metastatic phenotype by changing matrix structure. Finally, Adrian Harris, in your clinical studies is the association with TAMs and poor prognosis also strongly associated with IL10 production by most tumours?

Harris: We didn't look at IL10 on those tumours. We looked at quite a lot of other things. The sections are available so this could be done.

Mantovani: There are peculiar things in the matrix of tumours. For example, there is a fibronectin splice variant called EB. This is a domain of fibronectin conserved in evolution, so it must be important. It is very strongly expressed in the matrix of tumours. People are using it as a target to deliver drugs to tumours.

Balkwill: I want to return to the macrophages for a minute. At a recent meeting Adrian Harris said that 50% of all tumours are macrophages.

Harris: This was from old data. In some tumours it can be 50%.

Balkwill: Do all tumours have macrophages? Are macrophages a prerequisite for tumour growth?

Harris: We mainly looked at breast cancer. You can see macrophages in all tumours, but there is a great range in the amount you see. Often in early breast ductal cancer *in situ*, which is the preinvasive stage, you can often see massive swathes of macrophages around these ducts. Something is being secreted by these non-invasive tumours that is attracting macrophages in. No one has correlated the macrophage infiltrate with whether the cancers then become invasive or not. This needs to be done. There has been a suggestion that macrophages are a target for chemotherapy. Cisplatinum is toxic to macrophages in ovarian cancer, and this may contribute to the anticancer effect.

Balkwill: Someone once told me that when there is a response to chemotherapy in ovarian cancer activated macrophages can be found in the ascites.

Mantovani: There is an old story put forward by Steve Haskill suggesting that adriamycin activated macrophages.

Pollard: You would imagine that logically, if you start killing lots of tumour cells there will be a lot of antigen around. It is quite likely that this will be perceived as a wound. The macrophages would be newly recruited and then they would have a different phenotype, and perhaps even present those antigens at that stage. But when the tumour is normally growing and not dying. It inhibits this antigen presentation.

Gordon: When macrophages eat a lot of cells they tend to suppress immunity. They are not like DCs.

Richmond: I am interested in the microarray data. When cells are responding to a chemokine they are re-educating themselves continuously. They are activating a whole new set of transcriptional events which lead to changes in chemokine expression, amplification of this, as well as alterations in receptor expression. While the cells are migrating, they are changing their phenotype a bit as well.

Mantovani: One group of genes that came up from the expression profiling as being induced by CCL5 are genes involved in drug metabolism and carcinogen metabolism. If we have carcinogenesis data in targeted mice we should keep this in mind as an alternative possibility.

Oppenheim: Your proposed loop is very compelling, but I would like a bit more information on the CD4$^+$CD25$^+$ cell, which is a resting memory T cell. You point out that the TAMs generate PARC and MDC which are chemokines that particularly seem to be expressed by T$_{reg}$ immunosuppressive cells. This part of the loop makes a lot of sense, that a type II macrophage will then in turn have migrational effects on the T$_{reg}$ cells.

Mantovani: The gene chip data show that in some human tumours, such as in Hodgkin's disease, MDC/CCR22 comes out as a major gene expressed. This makes a lot of sense. The pathology of Hodgkin's is type II. In other tumours MDC has come up also. These are the data on which our scheme is based. The other piece of information is work from Daniele D'Ambrosio who has shown that T$_{reg}$ cells express CCL8 and CCL4. This makes sense.

The influence of CD25$^+$ cells on the generation of immunity to tumour cell lines in mice

Emma Jones, Denise Golgher*, Anna Katharina Simon, Michaela Dahm-Vicker, Gavin Screaton, Tim Elliott* and Awen Gallimore†[1]

*Nuffield Department of Medicine, Oxford University and *The Cancer Sciences Division, University of Southampton and †Medical Biochemistry and Immunology, University of Wales College of Medicine, UK*

Abstract. CD25$^+$ regulatory T cells comprise 5–10% of CD4$^+$ T cells in naïve mice and have been shown in several *in vivo* murine models to prevent the induction of autoimmune disease and inflammatory disease. Since T cells, which mediate autoimmunity, can through recognition of self-antigens also target tumour cells, it was postulated that CD25$^+$ regulatory cells would also inhibit the generation of immune responses to tumours. Depletion of these cells using CD25-specific monoclonal antibodies has indeed been shown to promote rejection of several transplantable murine tumour cell lines including melanoma, leukaemia and colorectal carcinoma. Results obtained using these models indicate that in the absence of regulatory cells, CD4$^+$ T cells mediate tumour immunity, although the precise mechanisms through which these cells result in tumour rejection have not yet been elucidated. The target antigens recognized by these CD4$^+$ T cells have also not yet been identified. Immunization of mice with tumour cells in the absence of CD25$^+$ regulatory cells does, however, induce immunity against a variety of different tumour cell lines indicating that the target antigen(s) are shared amongst tumours of distinct histological origins. Since CD25$^+$ regulatory cells have been identified in humans, the possibility that the cells inhibit immune responses to shared rejection antigens expressed by human tumours is worthy of investigation.

2004 Cancer and inflammation. Wiley, Chichester (Novartis Foundation Symposium 256) p 149–157

Accumulating evidence suggests that a population of CD4$^+$ T cells that constitutively express the interleukin (IL)2 receptor α chain (IL2Rα, CD25) function as regulatory cells capable of down-regulating immune responses to self-antigens. These cells, which represent 5–10% of peripheral CD4$^+$ T cells in

---

[1]This paper was presented at the symposium by Awen Gallimore to whom all correspondence should be addressed.

naïve mice, do not proliferate upon stimulation via their T cell receptor (TCR) but have been shown *in vitro* to suppress the activation of other T cells in an antigen-independent manner. Adoptive transfer of CD4$^+$CD25$^+$ regulatory T cells (T$_{reg}$) has been shown to prevent autoimmune and inflammatory disease in recipient mice.

Since tumour associated antigens are expressed both by tumours and normal tissue, we hypothesized that tumour immunity is also subject to inhibition by CD25$^+$ regulatory cells. We addressed this hypothesis by measuring induction of immunity to the melanoma cell line, B16F10, in C57BL/6 (B6) mice in the presence and absence of CD25$^+$ regulatory cells. Mice were injected with 1 mg of depleting antibodies specific for CD25 (PC61) or isotype control antibodies (GL113) three and one days prior to injection with the melanoma cell line, B16F10. Monitoring of tumour growth revealed that tumours grew more slowly, if at all, in mice injected with PC61 when compared to mice injected with GL113. In several similar experiments the proportion of PC61-treated mice remaining tumour free ranged from 30–60% whereas all control mice developed tumours. Those mice that remained tumour-free after challenge with B16F10 were re-challenged with a second inoculum of tumour cells approximately 8–12 weeks later. Approximately 70% of these mice resisted a second tumour challenge. Mice that had been treated with anti-CD25 monoclonal antibodies (mAbs) but which had not been previously inoculated with tumour cells were not able to reject tumours. To determine which immune subset was responsible for mediating long-term immunity to B16F10, purified CD8$^+$ and CD4$^+$ T cells, spleen cell eluates consisting of CD4-depleted or CD8-depleted spleen cells and serum recovered from tumour-free mice were injected into naïve, B6 mice. These mice were subsequently injected with B16F10 cells and tumour growth was monitored. Tumours grew in mice receiving CD8$^+$ T cells, spleen cells depleted of CD4$^+$ T cells or serum. Tumours did not develop in mice receiving CD4$^+$ T cells or CD8-depleted spleen cells thereby indicating that tumour immunity was mediated by CD4$^+$ T cells.

Next, we investigated whether or not T cell responses to melanocyte differentiation antigens were induced in tumour-free mice that had been treated with CD25-specific antibodies. Melanocyte antigens such as gp100, tyrosinase-related protein 1, tyrosinase-related protein 2, Melan-A and tyrosinase, which participate in melanin production, are expressed both by melanocytes and melanoma cells. In order to determine whether or not the tumour-free mice described above exhibited immunity to any of these antigens, we infected tumour-free mice with recombinant vaccinia viruses (rVVs) expressing either of these proteins. Reduction of viral titre in tumour-free mice compared to tumour-bearing or naïve control mice was taken as evidence of immunity to the recombinant proteins. A rVV expressing the irrelevant antigen, influenza nucleoprotein was used to control for antigen-specific elimination of rVVs expressing melanocyte antigens. A determination of virus titres indicated that 5/6 anti-CD25 mAb-treated tumour-free mice, but not naïve mice (0/3) or control mAb-treated, tumour-bearing mice (0/3) were partially protected against infection with rVV expressing the melanocyte antigen, tyrosinase but not rVV's expressing gp100, Trp-1, Trp-2, Melan-A or influenza nucleoprotein. Some mice treated with CD25-specific antibodies and which were tumour immune demonstrated evidence of depigmentation characterized by the appearance of white fur. All of these mice (9) were protected against infection with rVV expressing tyrosinase. These results indicate that inoculation of mice depleted of CD25$^+$ regulatory cells with melanoma cells can result in the induction of tyrosinase-specific autoreactivity. The data do not prove that T cells specific for melanocyte antigens are uncovered as a direct consequence of depletion of CD25$^+$ regulatory cells since these responses may represent downstream events of tumour destruction. In this case, CD25$^+$ regulatory cells may uncover antigen-non-specific immune responses or other antigen-specific responses important for tumour rejection. Indeed, we have observed that B16 tumours grow more quickly in B6 rag-deficient mice inoculated with purified CD4$^+$CD25$^+$ cells compared to B6 rag-deficient mice inoculated with no cells. Since B6 rag-deficient mice lack T and B cells, these data imply that CD25$^+$ regulatory cells inhibit antigen-non-specific immune responses. We are currently investigating this possibility.

In order to determine whether or not CD25$^+$ regulatory cells influence the development of immunity to other tumour cell lines in other strains of mice, we tested whether or not depletion of CD25$^+$ regulatory cells would facilitate the development of immunity to the colon carcinoma, CT26 in Balb/c mice. These experiments were carried out in collaboration with Dr Denise Golgher and Professor Tim Elliott at the University of Southampton. Balb/c mice were injected with CD25-specific antibodies or isotype-matched control antibodies prior to inoculation with CT26 cells. Unlike control mice, which were unable to reject the tumour challenge, 5/8 mice depleted of CD25$^+$ regulatory cells remained tumour free. Rejection of CT26 was accompanied by the development of lasting immunity to CT26 since the majority of mice were able to reject a second challenge of CT26 cells inoculated several weeks later. In order to determine the correlates of tumour immunity, we purified CD4$^+$ and CD8$^+$ T cells from CT26-immune mice and a control group of tumour-bearing mice and injected these into SCID recipient mice which were subsequently challenged with CT26. Both CD4$^+$ and CD8$^+$ cells from CT26-immune mice were able to mediate rejection of CT26 in recipient SCID mice. Interestingly CD8$^+$ T cells from tumour bearing mice were also able to mediate tumour rejection upon transfer into SCID mice. These results suggest that in the presence of CD25$^+$ regulatory cells, CT26 cells can stimulate CD8$^+$ but not CD4$^+$ T cell responses capable of tumour-rejection whereas in the absence of CD25⁺ cells, both CD4⁺ and CD8⁺ T cells can mediate tumour rejection.

To determine whether or not the target antigen(s) of immunity induced by CT26 in the absence of CD25⁺ regulatory cells were tumour-specific we challenged CT26-immune mice with a related colon carcinoma C26, a renal carcinoma RENCA, and the B cell lymphomas A20 and BCL-1. Surprisingly, CT26-immune mice were able to reject all of these tumours. These observations indicate that the target antigens of immune responses generated in CT26-immune mice depleted of CD25⁺ cells are shared tumour rejection antigens whose expression is not confined to tissue of colorectal origin. The nature of these antigen(s) is unknown. In experiments designed to determine which T cell subsets were responsible for the observed tumour cross-protection, we found that both CD4⁺ and CD8⁺ T cells were required. Since such cross-protection has rarely been observed in other models of transplanted murine tumours, it is tempting to speculate that absence of CD25⁺ regulatory cell activity uncovers immune responses to shared tumour rejection antigens.

Finally, we examined whether or not increasing the inflammatory potential of the tumour cells would improve their immunogenicity *in vivo*. Drs Katja Simon and Gavin Screaton had already determined that expression of Fas ligand on the cell line B16F10 (B16FasL) induced an inflammatory response upon inoculation into B6 mice. Indeed, approximately 50% of mice injected with B16FasL remain tumour-free and tumour rejection is associated with an infiltration of neutrophils at the site of inoculation. In collaboration with Katja and Gavin we also found that these tumour free mice were able to reject a subsequent inoculum of untransfected B16 cells. Transfer of serum or purified T cells from these mice into naïve recipient B6 mice revealed that tumour immunity was mediated by antibodies. We have also performed experiments which show that long-term immunity can be achieved in at least 90% of mice immunized with B16FasL in the absence of CD25⁺ regulatory cells. The superior protection against tumour growth obtained using the latter protocol could reflect induction of both an antibody response and a CD4⁺ T cell response capable of rejecting B16. Alternatively, depletion of CD25⁺ regulatory cells may enhance the antibody response induced by B16FasL. These possibilities are currently under investigation.

Acknowledgement

This work is supported by The Wellcome Trust.

DISCUSSION

Oppenheim: I have a specific question, concerning why in the RAG knockout mouse T$_{reg}$ administration seems to help the tumour growth. There are no T cells, NK cells or B cells, but there are macrophages. Do your T_{reg} cells make any products at all, such as TGFβ?

Gallimore: We haven't looked to see if they make TGFβ. As far as TGFβ is concerned, all we have done is neutralized it and found that the tumours don't grow very well. But there is still a missing link between that effect and whether the T_{reg} cells are responsible for that. The plan is to do some histology over sequential time points to see what cells are getting to the sites of tumour inoculation and what cytokines they make.

Brennan: My question concerns the kinetics of the depletion. I assume that the CD25+ regulatory cells come back. Have you done an experiment where you have challenged the mice after the cells have reappeared? It would be interesting to see whether the role of the regulatory cell is in preventing the priming of the immune response initially, or in the effector phase.

Gallimore: At the beginning we weren't even sure whether these antibodies were depleting or not. We tracked their presence or absence in the serum over time after injection, and also what was happening to the CD25+ cells. What happens is that, using our protocol, the antibody sticks around for about 21 d. From day 22 onwards, the CD25+ cells come trickling back. The experiment we did took a set of mice that had no detectable antibody in their serum and very few CD25+ cells in the periphery. Then we injected them with tumour cells. We still see the effect of suppression of tumour growth.

Brennan: If you challenged them when they are depleted, how long lived is that protection? Very clearly, they can mount an immune response in the absence of the CD25+ T_{reg} cells, but what happens when they are fully reconstituted as opposed to trickling back?

Gallimore: I am not sure. We do the second challenge months later, and those tumours are rejected. Perhaps you are suggesting that they have to come back to a certain level before they are effective again, and are they fully functional as soon as they come back? I'm not sure about this.

Blankenstein: Usually, rejection antigens are tumour-specific in mice. In your experiments, depletion of regulatory CD25+ T cells led to cross-protection against a variety of other tumours. Therefore, your results are very surprising. It would be interesting to see whether this cross-protection is antigen specific. In case of the tumour line CT26, a dominant antigen encoded by an endogenous retrovirus is shared with other tumour cell lines. Do you have evidence that there are some shared antigens such as this endogenous retrovirus which are responsible for cross-protection?

Gallimore: When we deplete the regulatory cells there may be antigen non-specific events that promote the generation of an antigen-specific immune response that is then responsible for clearing the secondary tumours. This is my interpretation. In terms of the retrovirus, there is an immunodominant epitope expressed by the envelope of a retrovirus, gp70. The peptide is a CD8 epitope called AH1. If we immunize mice with CT26 engineered to express GM-CSF, we see a very strong response to the AH1 peptide. Denise Golgher has by now looked at CTL responses to the AH1 peptide and finds that they are much stronger in the CT26-GM vaccinated mice than in mice treated with CD25-specific antibodies and which reject untransfected CT26. The retroviral antigen could be responsible for the cross-protecting immunity, but it is highly unlikely given that CT26-GM does not promote cross-protective immunity.

Blankenstein: This brings me to my second question, which concerns autoimmunity in this system. I am aware that it is difficult to study this because you don't know how to analyse clear signs of autoimmunity. Your experiment reminds me of Houghton's work. With melanoma-associated antigens Trp1 and Trp2, he has shown that one antigen induces antibody-mediated rejection, and the other cell-mediated tumour rejection. Signs of autoimmunity (vitiligo) are seen only during antibody-mediated rejection. Since you can transfer immunity with serum, do you also see autoimmunity in these mice?

Gallimore: No. Occasionally we see a very limited depigmentation around the site of tumour inoculation. There is a white patch.

Blankenstein: All the melanoma antigens that you listed are intracellular antigens. They should not be accessible to antibodies. By which mechanism do the antibodies directed against intracellular antigen induce autoimmunity?

Gallimore: We are not saying that the antibody is specific to those antigens. We have no evidence for that. It could be that the antibodies recognize proteins other than melanocyte antigens. The serum will stain B16 cells, other murine melanoma cells and human melanoma cells.

Gordon: Does it stain normal melanocytes?

Gallimore: I don't know.

Mantovani: Your finding in the Rag was provocative. Has anyone tested the role of $CD25^+$ cells in the control of conventional inflammatory reactions? For example, simple carrageenan-induced oedema?

Gallimore: Not to my knowledge.

Mantovani: Are we dealing with a control mechanism of inflammation, and then quite separately of specific immunity?

Gallimore: We are hoping to look at this in Cardiff using murine models of peritonitis. Fiona Powrie who works on a model of inflammatory bowel disease finds that $CD25^+$ cells prevent development of this disease in mice. I think she is dissecting what the relative effect of these cells is on the innate versus the acquired immune response. I think she also has some evidence that $CD25^+$ cells don't just suppress antigen-specific responses.

Hermans: Have you ruled the possibility that $CD4^+$ NK T cells could have a role? NKT cells have also been shown to have regulatory activity in some tumour models (Moodycliffe et al 2000). They can express CD25 as well (van der Vliet et al 2000). Do you know whether these cells are in any way involved in the anti-tumour responses you observe with anti-CD25 treatment?

Gallimore: We don't. We have done very little work on NK cells.

Mantovani: Has depletion of CD25+ cells any potential in a therapeutic setting?

Gallimore: We haven't had any success with B16. Experiments with CT26 look better. We wait until we see the tumours, give anti-CD25 antibodies, and look for tumour regression. In the one experiment performed by Denise Golgher, tumours disappeared in 2 out of 5 mice.

Caux: Going back to the cross-protection issue, have you looked for this in the B16 model?

Gallimore: No. We need to get hold of other tumour cell lines that will grow in B16 mice.

Blankenstein: It has been shown that Fas-ligand expression by tumour cells can induce tumour rejection. Granulocytes were suggested to be responsible for tumour rejection, since Fas-ligand appears to activate them. Are the granulocytes alone sufficient for the rejection of Fas ligand-expressing cells?

Gallimore: We haven't done this experiment. We also don't know whether the neutrophils are essential for the development of the long-term immunity. This is something that could be tested quite simply both in the presence and absence of regulatory cells. We need an anti-neutrophil antibody.

Richmond: What was the route of injection you used?

Gallimore: Tumours were inoculated subcutaneously.

Richmond: Did the size of the inoculum have any effect on the effectiveness of anti-CD25 antibody treatment?

Gallimore: Yes. We have to make sure that every control mouse gets a tumour. But with the melanoma cell line B16 we routinely inject a fairly low number of cells. They grow very fast. In most experiments we use 2×10^4 cells. These numbers are meaningless, in a way: different B16 lines grow very differently. If we increase the immunogenicity of the tumour we can get better rejection. With CT26 we can get away with injecting far more cells and still see an effect.

Richmond: Which cell line did you start with for B16?

Gallimore: It was F10. We have five different F10 lines from different labs and they are all quite quirky.

Oppenheim: As the animals recover and survive, do they recover some of their T regulatory cells with time? Do they develop any other autoimmune symptoms beside the vitiligo?

Gallimore: They seem to get their regulatory cells back. If we give the antibodies, wait three months and then challenge with tumour cells, the tumours will grow. With autoimmunity, occasionally we see depigmentation in the B16 mice, but this is quite rare. This is not something we would expect to see on a per-experiment basis and is probably something to do with the animal house. With BALB/c mice, we thought that since they are more susceptible to autoimmunity that we would see some pathology. We took the mice and challenged them with every single tumour to really push the response. We have done a bit of histology but we can't really see any autoimmune manifestations. But we are not really sure what we are looking for.

Oppenheim: Is the recovery or reappearance of the T_{reg} cells thymic dependent? You hinted that it might be peripheral.

Gallimore: This is really on the basis of other people's studies, where there is evidence that they can develop in the periphery. We have only once thymectomized mice and treated them with anti-CD25 antibodies. After a few months we did see quite a few CD25 cells back in the periphery, which seemed to have regulatory potential. This was however our first attempt at thymectomy!

Brennan: Have you tried looking the other way round? People interested in inducing regulatory cells, such as David Wraith, give autoantigens through the nasal route. Can you induce a non-rejected state by enhancing regulatory cells in your mice? Rather than depleting them, can you demonstrate that they are part and parcel of enhancing the tumorigenicity?

Gallimore: One way of addressing this experimentally might be to use non-depleting CD4-specific antibodies that can, using certain protocols, induce $CD25^+$ regulatory cells that inhibit graft rejection. In the same way, induction of $CD25^+$ regulatory cells might promote tumour growth.

Brennan: You can induce regulatory cells by giving autoantigens through intranasal routes, or feeding them in the diet. It is not clear whether we are inducing the same kinds of cells, or whether they are specific or non-specific.

Gallimore: They are non-specific in *in vitro* assays, but there is no real evidence that they are non-specific *in vivo*.

Brennan: Although they are specific to the antigen what they afford is a widespread regulatory effect against autoantigens within that tissue. As long as you direct them against an autoantigen directed, for example, in joint tissue, it affords some protection against a T cell response that is mounted against an autoantigen within the joint environment.

Gallimore: We have done some experiments in which we have transfected B16 cells with foreign antigens. The idea was that we would then see immune responses to these antigens due to depletion of the regulatory cells. However, we have seen no evidence of this. This sort of experimental system has a few flaws thereby making it difficult to determine whether or not regulatory cells inhibit responses in an antigen-specific or non-specific fashion.

Gordon: Is it possible to have a non-transplantation model and then try to deplete T regulatory cells?

Gallimore: We have collaborated with Pamela Ohashi who has produced mice named RIP-Tag2 which develop spontaneous tumours. We saw no effect on tumour growth when these mice were given CD25-specific antibodies. These mice develop tumours within weeks of life so I am not sure whether it is fair to address the issue within this experimental model. Another experiment we have in progress at the moment is designed to determine whether or not depletion of CD25⁺ cells inhibits tumour growth in mice injected with methylcholanthrene. This experiment is being carried out in collaboration with Maries Van den Broek. So far, 6 out of 9 control mice and 2 out of 9 mice treated with CD25-specific antibodies have developed tumours.

Richmond: One model that might work nicely is the tyrosinase promoter-*ras* transgenic mouse on a p16 knockout background. This is on a tetracycline-inducible promoter, so you could give your antibody to CD25 and then give the mice the deoxycytidine.

Blankenstein: I have a general question concerning anti-CD25 antibodies. The original motivation behind their use was to cure autoimmunity. Anti-CD25 antibodies are even used in the clinic to cure some forms of autoimmune disease. This creates quite a paradox. I ask myself, how good is CD25 as a marker? Which fraction of regulatory T cells are depleted by anti-CD25 antibodies?

Gallimore: For us, the situation is simpler. We take a naïve mouse. In this situation the CD25⁺ cells have regulatory activity, the others don't. As soon as mice have been challenged with antigen the populations become much more heterogeneous. It becomes difficult to separate them.

Blankenstein: You probably can't use anti-CD25 antibodies at a later time point, because then T cells against the tumour are activated and should express CD25. You would probably deplete those T cells, which are needed to eradicate the tumour.

Gallimore: It will be a balance between these things. We have looked at what happens to immune responses to viruses if mice are injected with anti-CD25 antibodies. It is surprising that T cell responses are not particularly affected by the presence of the antibodies. This work was done using influenza and vaccinia viruses.

Gordon: I think there will be a flood of new markers. We will have to wait and see whether they are useful or not.

References

Moodycliffe AM, Nghiem D, Clydesdale G, Ullrich SE 2000 Immune suppression and skin cancer development: regulation by NKT cells. Nat Immunol 1:521–525

van Der Vliet HJ, Nishi N, de Gruijl TD et al 2000 Human natural killer T cells acquire a memory-activated phenotype before birth. Blood 95:2440–2442

Macrophages: modulators of breast cancer progression

Elaine Y. Lin and Jeffrey W. Pollard[1]

Center for the Study of Reproductive Biology and Women's Health, Departments of Developmental and Molecular Biology and Obstetrics, Gynecology and Women's Health, Albert Einstein College of Medicine, 1300 Morris Park, New York, NY 10461, USA

Abstract. In many solid tumour types the abundance of tumour associated macrophages (TAMs) is correlated with poor prognosis. Macrophages are recruited through the local expression of chemoattractants such as colony stimulating factor 1 (CSF-1) and macrophage chemoattractant protein 1. Over-expression of both of these factors is correlated with poor prognosis in a variety of tumours. Macrophages also play an important physiological role in the development and function of many tissues ranging from the brain to the mammary gland. Thus we hypothesized that TAMs are recruited to tumours through the expression of potent chemoattractants and in this site their normal trophic functions are subverted to promote tumour progression and metastasis. To test this hypothesis we crossed mice deficient in macrophages owing to being homozygous for a null mutation in the CSF-1 gene with mice pre-disposed to mammary cancer due to the epithelial restricted expression of the polyoma middle T oncoprotein. The absence of macrophages did not change the incidence or growth of the primary tumour but decreased its rate of progression and inhibited metastasis. These data are explicable through the known macrophage functions in matrix re-modelling, angiogenesis and stimulation of tumour growth and motility through the synthesis of growth and chemotactic factors. Interestingly, these functions are also normally found in wound healing or pathologically during chronic inflammation. This supports the notion that tumours are 'wounds that never heal' and suggests that chronic inflammation through persistent infection or by other means might be an important co-factor in the genesis and promotion of tumours. Macrophages might therefore be important targets for cancer therapies.

2004 Cancer and inflammation. Wiley, Chichester (Novartis Foundation Symposium 256) p 158–172

The discovery of genes capable of transforming cells to a malignant phenotype in culture and causing tumours in mice, led to a focus upon intrinsic events in epithelial cells that cause cancer. The need for more than one oncogene as well as

---

[1]This paper was presented at the symposium by Jeffrey W. Pollard to whom correspondence should be addressed.

the loss of tumour suppressor genes to give complete neoplastic transformation also was intellectually compatible with epidemiological studies with human populations that suggested multiple events over long periods required for the generation of metastatic disease (Cairns 1981, Wu & Pandolfi 2001). The power of these studies on cancer causing mutations however, led to the down play of the study of host factors that influence the development of tumours *in vivo*. This balance is now being redressed with the appreciation that epithelial tumours develop in a complex stroma and are dependent upon many of these cells for their nutrition, invasiveness and the expression of a full neoplastic phenotype. Indeed, there has been considerable recent interest in angiogenesis, a process required for tumour development and which acts as a target for chemotherapy (Carmeliet & Jain 2001, Giordano & Johnson 2001, Kerbel 2000). The notion that the surrounding stroma plays an important role in carcinogenesis is of course not new. In fact, early views on the formation of cancers suggested they were the result of chronic inflammation and that tumours were 'wounds that never heal' (Balkwill & Mantovani 2001). More recently there has been increasing epidemiological evidence that inflammation, often as a result of incipient infection, can act as a promoter of tumorigenesis (Balkwill & Mantovani 2001, Coussens & Werb 2001, 2002). Furthermore, experimental studies in mice have supported this view by showing that cells of the innate immune system that play a central role in inflammatory responses also play crucial roles in tumour progression (Coussens et al 2000, 1999, Lin et al 2002, 2001). This review will concentrate upon one class of these cells, the macrophage.

Tumour-associated macrophages

Macrophages are found in many solid tumours where they are known as tumour-associated macrophages (TAMs). Their function has been a point of controversy since these cells are multifunctional and their activity context dependent (Leek & Harris 2002). They have been considered to have anti-tumour actions through their immune functions while more recently their pro-tumour roles have become more evident. Of course, their most studied function is in immunity where they act as central signalling cells of the innate immune response (Nathan 2002). They are instrumental in the switch from innate to adaptive immunity and they also function as antigen presenting cells to effector T cells leading to pathogen eradication and transplant rejection. This latter role led to the supposition that TAMS might reject the 'foreign' tumour tissue. However, the evidence for tumour rejection has often been based upon xenograft experiments where the tumour is indeed foreign and thus these data may simply reflect a 'host versus graft' response. Nevertheless, the search for true tumour antigens in naturally occurring tumours has been pursued vigorously with several having been identified. However, even when *bona fide* tumour antigens have been discovered, such as the unique MUC1 epitope in human breast cancers, there is little evidence for a vigorous host response to these antigens (Ohm & Carbone 2001, Taylor-Papadimitriou et al 2002). Instead recent data suggest that many cancers, particularly breast cancers, suppress both systemic and local immunity (Almand et al 2000, Igney & Krammer 2002).

The focus upon the immunological functions of macrophages has often resulted in their remodelling and trophic functions being overlooked (Scholl et al 1993, Mantovani 1994, O'Sullivan & Lewis 1994). These functions are effected through their production of growth and angiogenic factors, matrix metalloproteinases and by their phagocytic activity (Nathan 2002, Witty et al 1995, Martin 1997, Witte & Barbul 2002). Studies in mice that are macrophage-deficient due to their being homozygous for a null mutation, $Csf1^{op}$, in the gene for the major mononuclear phagocyte growth, viability and chemotactic factor, colony-stimulating factor 1 (CSF1, or macrophage-CSF) have emphasized these trophic roles for macrophages during development. Thus many of their phenotypes can be explained by the absence of macrophage trophic and remodelling functions (Pollard & Stanley 1996). Most prominent is their osteopetrosis that is due to the absence of osteoclasts, cells of the mononuclear phagocytic lineage that remodel bone. But there are also many other phenotypes especially those that impact reproduction that indicate macrophages have essential roles in the development of tissue as diverse as the brain and mammary gland (Pollard & Stanley 1996, Pollard 1997, Gouon-Evans et al 2002). In the mammary gland, for example, macrophages enhance branching morphogenesis and in their absence the mammary gland is an atropic, poorly branched structure (Gouon-Evans et al 2000). Given these roles in normal development, it seems probable that TAMs perform some of these functions with the consequent promotion of tumour progression and metastasis. Consistent with this concept a recent survey of the clinical data on TAMs in breast cancer, concluded, based upon five independent studies, that there is a strong correlation between TAM density and poor prognosis. Similar results were found for non-small cell lung (Shijubo et al 2003), bladder and ovarian cancers while for prostate, lung and brain cancers conflicting data have been obtained (Bingle et al 2002).

Tissue macrophages and TAMs are derived from circulating monocytes that are recruited into tissues in response to chemoattractants (Mantovani 1994). These include monocyte chemoattractant proteins (MCP-1–5), macrophage inflammatory protein 1α (MIP1α), macrophage migration inhibition factor (MIF) and CSF-1 (Baggiolini et al 1997). There is substantial clinical evidence in humans that the overexpression of CSF-1 is associated with poor prognosis in breast, ovarian, endometrial and prostatic cancers (Kacinski 1995, 1997, Sapi & Kacinski 1999, Scholl et al 1994, Smith et al 1995, Ide et al 2002). In breast cancer, CSF-1 expression was especially prominent at the invasive edge of the tumour and its overexpression was correlated with leukocytic infiltration (Scholl et al 1994, Tang et al 1992). Recent data have shown that MCP-1 overexpression is also strongly correlated with poor prognosis in breast cancers (Saji et al 2001) and that MCP-1 is a major determinant of macrophage content in many tumours (Wong et al 1998).

Macrophages enhance tumour progression and metastatic rate in mouse models of cancer

In light of these clinical observations that show correlations between over-expression of macrophage chemoattractants and macrophage density in tumours with poor prognosis and our studies on the role of macrophages during normal development mentioned above, we suggested that TAMs provide trophic support to the developing tumour that enhances its progression and facilitates its metastatic capability (Lin et al 2001). In order to test the hypothesis, we crossed mice with a macrophage deficiency caused by homozygosity of the *Csf1op* mutation, with those pre-disposed to mammary cancers because of the expression of the polyoma middle T oncoprotein (PyMT) in the mammary epithelium (Lin et al 2001). This mouse model of breast cancer was chosen because of its high penetrance with 100% of the mice succumbing by four to six weeks of age and with a high percentage of mice exhibiting lung metastasis by 22 weeks of age. Furthermore, we have defined in detail, both at the histological and biochemical level, the progressive transitions that the tumours take to their eventual destiny as late carcinomas. This has allowed the quantitative measurement of tumour progression in this model (Lin et al 2002, 2001).

The removal of macrophages from this tumour model did not result in a reduction in the incidence or growth of the primary tumour. However, it did slow the rate of tumour progression to the most malignant stages and reduced the rate of metastasis to virtually zero (Lin et al 2001). To ensure that these effects were intrinsic to the mammary gland, we used transgenic technology to re-introduce CSF-1 specifically into the mammary epithelium. This resulted in a dramatic recruitment of macrophages to the tumour and a restoration of metastasis and an enhancement of the rate of progression to the late carcinoma stage (Fig 1). Furthermore, over-expression of CSF-1 in the wild type mammary gland recruited macrophages earlier, enhanced tumour progression and doubled the rate of metastasis (Lin et al 2001). These data strongly argue for a positive impact of macrophages upon tumour progression and lead to the conclusion that these cells potentiate the metastatic rate. These data also suggest that the clinical correlations observed between TAMs, CSF-1 overexpression and poor prognosis have a causal basis.

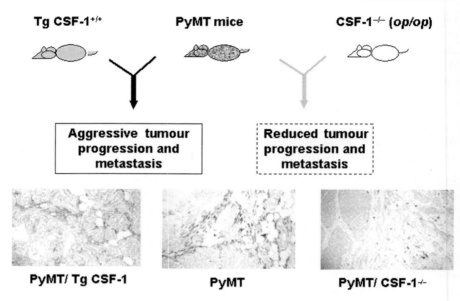

FIG. 1.   Macrophage infiltration affects tumour progression and metastasis. Mice with a predisposition to mammary cancers due to the mammary epithelial restricted expression of the polyoma middle T oncoprotein (PyMT mice) were crossed with macrophage deficient mice (*Csf1*$^{-/-}$ *op/op*). This resulted in fewer TAMs shown in the lower histological panel (dark staining cells) and a reduction in the rate of tumour progression and metastasis. Reintroduction of CSF-1 into the null background by transgenic means (Tg *Csf*$^{+/+}$) restored the TAM density and resulted in aggressive tumours with high metastatic potential. Overexpression of CSF-1 using the same technology in the wild-type mice resulted in enhanced TAM recruitment and an acceleration of the rate of tumour progression and metastasis. The bottom panels show histological sections of late stage PyMT tumours from the different genotypes that had been immunostained for macrophages using anti-F4/80 antibody. These cells are shown as dark cells against the lightly haematoxylin-stained background.

Other studies suggest that these results can be generalized to other tumour types. Treatment of mice transplanted with either human embryonic or colon tumours with CSF-1 antisense oligonucleotides directed specifically to mouse CSF-1 suppressed growth of the tumours and blocked metastasis. These results suggest that inhibition of CSF-1 synthesis blocked tumour recruitment of macrophages and consequently reduced tumour progression (Aharinejad et al 2002). In another study in which lung metastases produced from i.v. injected tumour cells were enhanced by the presence of distant primary tumours, it was demonstrated that TAMs provided the matrix metalloprotease, MMP9, that was involved in the up-regulation of VEGF locally in the lung that in turn was required for lung metastases to occur (Hiratsuka et al 2002).

Molecular mechanisms of TAM action

The clinical and experimental data presented above provide a compelling argument that at least in some tumours, TAMs play an important role in tumour progression and metastasis. TAMs are able to produce many molecules that can influence these processes including proteases, angiogenic and growth factors/chemokines and immunosuppressive molecules (Fig. 2) (Leek & Harris 2002, Mantovani et al 1992). A study by Harris and co-workers (Leek et al 1996) showed macrophage hot-spots in the tumour bed correlated with high levels of angiogenesis and worsened prognosis and relapse-free survival. These macrophages tended to cluster at avascular hypoxic areas (Leek et al 2000, Lewis et al 2000). This led to the hypothesis that hypoxia in these areas causes recruitment of macrophages and the expression within the cells of the hypoxia-induced transcription factor (HIF-2α). Pro-angiogenic cytokines such as VEGF are targets of this transcription factor (Leek & Harris 2002). VEGF is also a chemoattractant to monocytes perhaps reinforcing the recruitment of TAMs to the avascular areas (Barleon et al 1996). TAMs also synthesize many other angiogenic factors including the angiopoietins, FGF2, thymidine phosphorylase and TNFα (Leek & Harris 2002). In addition, TAMs produce many proteases such as uPA and MMP9, whose action may release angiogenic factors in an active form from the extracellular matrix (Leek & Harris 2002). This process may be regulated

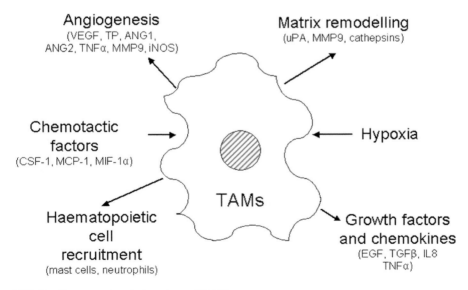

FIG. 2.   Pro-tumorigenic functions of TAMs.

by TGFβ1 whose concentration is elevated in late-stage breast cancers (Hildenbrand 1998). These actions of TAMs on angiogenesis are supported by the experiments using CSF-1 antisense mentioned above that reduced tumour angiogenesis and inhibited tumour progression and metastasis (Aharinejad et al 2002). Thus TAMs appear to play a major role in angiogenesis in cancers.

TAMs are also known to produce proteases and a variety of growth factors (Fig 2) (Leek & Harris 2002). Interestingly in our studies of the progression of PyMT-induced mammary tumours in the mouse model, it was apparent from histological observations that TAMs were recruited abundantly to the tumour stroma at the malignant transition (Lin et al 2001). It was also noticeable beginning at this stage that there were leukocytic infiltrates involving TAMs and other haematopoietic cells that were associated with a breakdown of the basement membrane surrounding the tumour. Thus TAMs might cause the focal breakdown of basement membranes through their production of proteases. TAMs would also enhance the formation of vessels at these invasion sites. Furthermore, they would provide chemotactic factors and growth factors, such as EGF, that would stimulate the motility and proliferation of tumour cells locally. These actions of TAMs in tumours is reminiscent of their role during wound healing suggesting that tumours subvert normal inflammatory processes to their own ends (Mantovani et al 1992). During wound healing macrophages act to organize and recruit other haematopoietic cells (Crowther et al 2001). Substantial experimental data exist suggesting that haematopoietic cells other than macrophages, particularly mast cells, play a major role in tumour progression through their production of angiogenic factors and proteases (Coussens et al 2000, 1999). Thus TAMs might have secondary effects through other cells. In wound healing all these haematopoietic cells act to produce both chemotactic and growth factors for epithelial cells that stimulate their migration and proliferation to effect wound closure and healing. However, in contrast to wound healing where the epithelial cells retain positional identity and cease proliferation appropriately, these actions in the tumour would stimulate continuous proliferation and allow escape of the epithelial tumour cells that have now lost their positional identity through intrinsic mutations. In addition these actions would select for those tumour cells that are capable of movement out of the tumour and into the vasculature.

In another but not mutually exclusive scenario, persistent infection could result in the chronic inflammation that acts to aid tumour development. During chronic inflammation such as in that observed in arthritis, macrophages are the major producers of pro-inflammatory cytokines that result in the pathologies that include matrix erosion, angiogenesis and enhanced cellular proliferation (Campbell et al 2000). These processes could act as promoters of the original carcinogenic events in the epithelial cells. Alternatively, macrophages might themselves initiate cancer since in such a situation through their excessive production of oxygen radicals, they might be mutagenic to epithelial cells (Fulton et al 1984, Kim et al 2003). Significantly, at least 15% of tumours have their origin in persistent infections (Coussens & Werb 2001, Kuper et al 2000).

Conclusions

There is substantial evidence that the tumour stroma plays a major role in the modulation of a tumour's phenotype. Many cells contribute to this including adipocytes, fibroblasts and immune cells. These immune cells lend their normal functions such as those involved in tissue remodelling and angiogenesis to the tumour to enhance its growth. Prominent among the immune cells are macrophages whose abundance correlates with poor prognosis in many tumour types. Several experiments in mice have confirmed the importance of these cells in tumour progression and metastasis.

Taken together all these data suggest that modulation of macrophage functions, particularly their inflammatory roles, might have a major therapeutic impact in a wide range of cancers. It is notable that intervention studies both in mice and humans with non-steroidal anti-inflammatory drugs (NSAIDs) that inhibit COX-2 activity reduced the risk of colon cancer. COX-2 is the major enzyme in the prostaglandin biosynthetic pathway and macrophages are significant producers of prostaglandins especially during inflammation (Ricchi et al 2003). It will be important to establish if these NSAIDs directly target macrophages since this would further establish a causal link between macrophage function and cancer progression and enhance the view that anti-macrophage therapeutics could be beneficial in cancer treatment and prevention (Bingle et al 2002).

References

Aharinejad S, Abraham D, Paulus P et al 2002 Colony-stimulating factor-1 antisense treatment suppresses growth of human tumor xenografts in mice. Cancer Res 62:5317–5324

Almand B, Resser JR, Lindman B et al 2000 Clinical significance of defective dendritic cell differentiation in cancer. Clin Cancer Res 6:1755–1766

Baggiolini M, Dewald B, Moser B 1997 Human chemokines: an update. Annu Rev Immunol 15:675–705

Balkwill F, Mantovani A 2001 Inflammation and cancer: back to Virchow? Lancet 357:539–545

Barleon B, Sozzani S, Zhou D, Weich HA, Mantovani A, Marme D 1996 Migration of human monocytes in response to vascular endothelial growth factor (VEGF) is mediated via the VEGF receptor flt-1. Blood 87:3336–3343

Bingle L, Brown NJ, Lewis CE 2002 The role of tumour-associated macrophages in tumour progression: implications for new anticancer therapies. J Pathol 196:254–265

Brannstrom M, Enskog A 2002 Leukocyte networks and ovulation. J Reprod Immunol 57: 47–60

Cairns J 1981 The origin of human cancers. Nature 289:353–357

Campbell IK, Rich MJ, Bischof RJ, Hamilton JA 2000 The colony-stimulating factors and collagen-induced arthritis: exacerbation of disease by M-CSF and G-CSF and requirement for endogenous M-CSF. J Leukoc Biol 68:144–150

Carmeliet P, Jain RK 2000 Angiogenesis in cancer and other diseases. Nature 407:249–257

Coussens LM, Werb Z 2001 Inflammatory cells and cancer: think different! J Exp Med 193: F23–F26

Coussens LM, Werb Z 2002 Inflammation and cancer. Nature 420:860–867

Coussens LM, Raymond WW, Bergers G et al 1999 Inflammatory mast cells up-regulate angiogenesis during squamous epithelial carcinogenesis. Genes Dev 13:1382–1397

Coussens LM, Tinkle CL, Hanahan D, Werb Z 2000 MMP-9 supplied by bone marrow-derived cells contributes to skin carcinogenesis. Cell 103:481–490

Crowther M, Brown NJ, Bishop RT, Lewis CE 2001 Microenvironmental influence on macrophage regulation of angiogenesis in wounds and malignant tumors. J Leukoc Biol 70:478–490

Fulton AM, Loveless SE, Heppner GH 1984 Mutagenic activity of tumor-associated macrophages in Salmonella typhimurium strains TA98 and TA 100. Cancer Res 44:4308–4311

Gillitzer R, Goebeler M 2001 Chemokines in cutaneous wound healing. J Leukoc Biol 69: 513–521

Giordano FJ, Johnson RS 2001 Angiogenesis: the role of the microenvironment in flipping the switch. Curr Opin Genet Dev 11:35–40

Gouon-Evans V, Rothenberg ME, Pollard JW 2000 Postnatal mammary gland development requires macrophages and eosinophils. Development 127:2269–2282

Gouon-Evans V, Lin EY, Pollard JW 2002 Requirement of macrophages and eosinophils and their cytokines/chemokines for mammary gland development. Breast Cancer Res 4:155

Hildenbrand R, Jansen C, Wolf G et al 1998 Transforming growth factor-beta stimulates urokinase expression in tumor-associated macrophages of the breast. Lab Invest 78:59–71

Hiratsuka S, Nakamura K, Iwai S et al 2002 MMP9 induction by vascular endothelial growth factor receptor-1 is involved in lung-specific metastasis. Cancer Cell 2:289–300

Ide H, Seligson DB, Memarzadeh S et al 2002 Expression of colony-stimulating factor 1 receptor during prostate development and prostate cancer progression. Proc Natl Acad Sci USA 99:14404–14409

Igney FH, Krammer PH 2002 Immune escape of tumors: apoptosis resistance and tumor counterattack. J Leukoc Biol 71:907–920

Kacinski BM 1995 CSF-1 and its receptor in ovarian, endometrial and breast cancer. Ann Med 27:79–85

Kacinski BM 1997 CSF-1 and its receptor in breast carcinomas and neoplasms of the female reproductive tract. Mol Reprod Dev 46:71–74

Kerbel RS 2000 Tumor angiogenesis: past, present and the near future. Carcinogenesis 21: 505–515

Kim HW, Murakami A, Williams MW, Ohigashi H 2003 Mutagenicity of reactive oxygen and nitrogen species as detected by co-culture of activated inflammatory leukocytes and AS52 cells. Carcinogenesis 24:235–241

Kuper H, Adami HO, Trichopoulos D 2000 Infections as a major preventable cause of human cancer. J Intern Med 248:171–183

Leek RD, Harris AL 2002 Tumor-associated macrophages in breast cancer. J Mammary Gland Biol Neoplasia 7:177–189

Leek RD, Lewis CE, Whitehouse R, Greenall M, Clarke J, Harris AL 1996 Association of macrophage infiltration with angiogenesis and prognosis in invasive breast carcinoma. Cancer Res 56:4625–4629

Leek RD, Hunt NC, Landers RJ, Lewis CE, Royds JA, Harris AL 2000 Macrophage infiltration is associated with VEGF and EGFR expression in breast cancer. J Pathol 190:430–436

Lewis JS, Landers RJ, Underwood JC, Harris AL, Lewis CE 2000 Expression of vascular endothelial growth factor by macrophages is up-regulated in poorly vascularized areas of breast carcinomas. J Pathol 192:150–158

Lin EY, Nguyen AV, Russell RG, Pollard JW 2001 Colony-stimulating factor 1 promotes progression of mammary tumors to malignancy. J Exp Med 193:727–740

Lin EY, Gouon-Evans V, Nguyen AV, Pollard JW 2002 The macrophage growth factor, CSF-1, in mammary gland development and tumor progression. J Mammary Gland Biol Neoplasia 7:147–162

Mantovani A 1994 Tumor-associated macrophages in neoplastic progression: a paradigm for the in vivo function of chemokines. Lab Invest 71:5–16

Mantovani A, Bottazzi B, Colotta F, Sozzani S, Ruco L 1992 The origin and function of tumor-associated macrophages. Immunol Today 13:265–270

Martin P 1997 Wound healing—aiming for perfect skin regeneration. Science 276:75–81

Nathan C 2002 Points of control in inflammation. Nature 420:846–852

Ohm JE, Carbone DP 2001 VEGF as a mediator of tumor-associated immunodeficiency. Immunol Res 23:263–272

O'Sullivan C, Lewis CE 1994 Tumour-associated leucocytes: friends or foes in breast carcinoma. J Pathol 172:229–235

Pollard JW 1997 Role of colony-stimulating factor-1 in reproduction and development. Mol Reprod Dev 46:54–60

Pollard JW, Stanley ER 1996 Pleiotropic roles for CSF-1 in development defined by the mouse mutation osteopetrotic (op). Adv Dev Biochem 4:153–193

Ricchi P, Zarrilli R, Di Palma A, Acquaviva AM 2003 Nonsteroidal anti-inflammatory drugs in colorectal cancer: from prevention to therapy. Br J Cancer 88:803–807

Sapi E, Kacinski BM 1999 The role of CSF-1 in normal and neoplastic breast physiology. Proc Soc Exp Biol Med 220:1–8

Saji H, Koike M, Yamori T et al 2001 Significant correlation of monocyte chemoattractant protein-1 expression with neovascularization and progression of breast carcinoma. Cancer 92:1085–1091

Scholl SM, Crocker P, Tang R, Pouillart P, Pollard JW 1993 Is colony stimulating factor-1 a key mediator in breast cancer invasion and metastasis? Mol Carcinog 7:207–211

Scholl SM, Pallud C, Beuvon F et al 1994 Anti-colony-stimulating factor-1 antibody staining in primary breast adenocarcinomas correlates with marked inflammatory cell infiltrates and prognosis. J Natl Canc Inst 86:120–126

Shijubo N, Kojima H, Nagata M et al 2003 Tumor angiogenesis of non-small cell lung cancer. Microsc Res Tech 60:186–198

Smith HO, Anderson PS, Kuo DYK et al 1995 The role of colony-stimulating factor 1 and its receptor in the etiopathogenesis of endometrial adenocarcinoma. Clin Canc Res 1: 313–325

Tang R, Beuvon F, Ojeda M, Mosseri V, Pouillart P, Scholl S 1992 M-CSF (Monocyte Colony Stimulating Factor) and M-CSF receptor expression by breast tumor cells: M-CSF mediated recruitment of tumour infiltrating monocytes? J Cell Biochem 50:350–3560

Taylor-Papadimitriou J, Burchell JM, Plunkett T et al 2002 MUC1 and the immunobiology of cancer. J Mammary Gland Biol Neoplasia 7:209–221

Witte MB, Barbul A 2002 Role of nitric oxide in wound repair. Am J Surg 183:406–412

Witty JP, Wright JH, Matrisian LM 1995 Matrix metalloproteinases are expressed during ductal and alveolar mamary morphogenesis, and misregulation of stromelysin-1 in transgenic mice induces unscheduled alveolar development. Mol Biol Cell 6:1287–1303

Wong MP, Cheung KN, Yuen ST et al 1998 Monocyte chemoattractant protein-1 (MCP-1) expression in primary lymphoepithelioma-like carcinomas (LELCs) of the lung. J Pathol 186:372–377

Wu X, Pandolfi PP 2001 Mouse models for multistep tumorigenesis. Trends Cell Biol
    11:S2–S9

DISCUSSION

Richmond: In your recent microarray work, have you looked at chemokine
receptor expression on those cells that happen to fall into the needles you use to
take your samples?

Pollard: Not yet. We only get 1000 cells. We have spent the last 18 months or so
working out array systems with Smart amplification which would give us linear
amplification. We have now done that and can amplify from about 200 cells in a
proportional way. At Einstein we have a microarray facility which we built
ourselves which has 27 000 genes per chip. We have started to array these cells.
For the macrophages we have to immunodeplete tumour cells, so we have pure
macrophage populations. We don't have any data on this yet. Our collaborators
have some data on motility and Rho-type proteins which are overexpressed in these
tumour cells.

Richmond: So you couldn't immunostain the cells.

Pollard: We could do, but there are so few of them. We haven't done this.

Feldmann: It seems that your progressive model is a wonderful way to look at
what stage angiogenesis begins. Have you had an opportunity to look at when the
angiogenic factors come up?

Pollard: Yes, it is a wonderful system. Elaine Lin has been working on this for
the last 18 months but she won't let me talk about it. One of the effects of the
absence of macrophages is that vasculogenesis is impaired, and the type of
vasculogenesis is different.

Rollins: There is a huge amount of information content in the Tet-inducible
CSF-1 part of the model that you described in the first part of your talk. What
you have been able to do is divorce the macrophage proliferative effects of CSF-1
from the recruitment effects. The reason I say this is that several years ago we made
a MMTV MCP-1 mouse. Our expectation was that we would see tons of
macrophages in mammary tissue, but we saw nothing. There is no way in that
mouse that you can get macrophages to go anywhere. The reason for this is that
the MMTV promoter is powerful enough, and there is enough expression in
salivary gland, mammary tissue, gonads, lung and kidney that there are
circulating levels of MCP-1. This either desensitizes receptors on the surface of
circulating monocytes or it physically cancels out the chemoattractant
concentration gradient, so nothing can migrate in response to MCP-1 in that
model. I would suggest that in your model, where you have Tet-inducible
MMTV driving CSF-1 that you are not going to get migration in response to
that. Instead, you will get expansion of monocytes and then secondarily probably
MCP-1 or something else is attracting those cells into the mammary tissue. It is

actually a very clean system in which you have shown that CSF-1 is either activating or expanding monocytes, but probably not attracting them.

Pollard: I think it is actually attracting them. This is a tight system. The only other tissue that seems to overexpress CSF-1 is the salivary gland.

Rollins: Presumably CSF-1 is getting into the circulation.

Pollard: None does; we have measured this. It is really a recruitment. I don't know whether this recruitment requires other factors as well. It may do. We feel that it is probably sufficient on its own. We have known for a long time, since our early reconstitution experiments with Richard Stanley, that when we injected CSF-1 locally it would recruit macrophages locally.

Rollins: In these reconstitutions, do you need CSF-1 in the marrow in order to see expansion?

Pollard: No. There are sufficient monocytes and their proliferative capacity is high enough that even though there is less than 1% of the normal level, we can get a normal level of macrophages. We can't exclude some small leakage, but the radioimmuno assays (RIAs) are sensitive to 1 pica mole and we can't detect anything with this RIA.

Harris: I liked your theory that you are recapitulating a normal developmental process. Is there a defect in the modelling of the breast, and if you look at your inducible CSF-1 model without the cancer can you restore that to normal?

Pollard: Yes. This is one of the situations where we can actually turn it on and off, stopping and restarting development.

Harris: Can you do this in any other tissues?

Pollard: No, that's the only one we have done it in so far. With the *Csf1^{op}/Csf1^{op}* mice what we find is that macrophages are not specification factors. But what they do is promote development and we think they are very important for branching morphogenesis. If you look developmentally you see a delay, but eventually the system catches up. Either there are enough macrophages (there are some in the *Csf1^{op}/Csf1^{op}* mouse) or other factors come along and replace them. In the bone, for example, in the *Csf1^{op}/Csf1^{op}* mouse the bone marrow cellularity is eventually restored and the cavity is moulded out. This turns out to be VEGF receptor signalling to a Flk receptor on osteoclast progenitors. Eventually it takes over from CSF-1 but it takes about six weeks as opposed to the normal two weeks. In the mammary gland the *Csf1^{op}/Csf1^{op}* mouse ends up with a ductal structure which fills the whole fat, but it is atrophic. The major thing is that there is a delay in the branching morphogenesis. It seems that macrophages are really promoting factors enabling the system to work at the right rate.

Harris: Can you switch it off again and stop tumours?

Pollard: We haven't done that experiment. These experiments are difficult because there are four alleles. If you just think about the number of mice needed to do this, it is a huge cohort.

Wilkins: I also like this idea that what is seen in the tumour is a recapitulation of the normal developmental mechanism. This allows us to understand the origins of this complex response as an evolutionary effect. Otherwise it is hard to imagine how these complex interactions could ever have evolved. It reminds me of the fact that certain vertebrates, such as sharks, are highly resistant to tumour induction. Has anyone looked at whether their organogenesis doesn't involve macrophage remodelling in the same way?

Pollard: Not me: sharks don't have mammary glands!

Mantovani: You have shown that macrophages play a crucial role in the transition from adenoma to carcinoma. Is this simply through the breaking of the basal membrane? Are they actually promoting a genetic event?

Pollard: This is a good question. It was the most perplexing of the results we saw. I suppose you could look at it either way. The macrophages could be producing proteases which allow remodelling of the matrix, and then this enhances the survival of more virulent genetic variants within the tumours. They could be producing apoptotic resistance factors or survival factors which would be promoting the selection of more aggressive variants. Or, since they do produce reactive oxygen, they could be making a mutagenic environment and causing an increased rate of mutation. We don't have any way of distinguishing among these particular events. Some people have suggested that the polyoma middle T model is so aggressive that secondary events are unnecessary, but this is not true.

Mantovani: I have another question concerning the proliferative effect of CSF-1. This is clearly different in rodents versus humans. Mature macrophages in the mouse proliferate to some extent, whereas it is very hard to see proliferating macrophages in humans. Are the macrophages in this system proliferating? A while ago we looked for proliferating macrophages in ovarian cancer, but we got negative results. There was something in Kaposi sarcoma, and I wonder whether there are proliferating macrophages in human tumours.

Harris: We have never seen them.

Pollard: In our case, one of the things we wanted to do at the beginning was to see what the proliferation rates were like. We were getting normal growth rates so we assumed that proliferation rates were similar. They were. We never got the impression that any of the stromal cells were really proliferating.

Blankenstein: Do you know when the tumour starts to make CSF-1?

Pollard: The epithelial cells make CSF-1 so it is seen right from the beginning, even in a normal mammary gland. There is a lot more later on, but then there are more epithelial cells. We don't know on a per cell basis whether more is being made. All we know is that CSF-1 is there all the time. We have now made a mouse where we have β gal under the CSF-1 promoter as a marker. We have always had problems with immunohistochemistry for CSF-1. Over two dozen antibodies have been made but none of them work for immunohistochemistry. The message level is quite low so *in situ* it is hard to see. We are just crossing this reporter mouse into this model to look at the expression patterns of CSF-1.

Thun: This is an interesting set of experiments and theory. Where do you go from here? Specifically, will you be looking at whether this is operative in lots of different models? How do you test its relevance to human cancers?

Pollard: I would like to think that tumour-associated macrophages have a general relevance to many solid tumours. We are testing this out in other breast cancer models. Is overexpression of CSF-1 a marker for many tumours? We have shown, in particular for endometrial cancer, that it is a good marker for poor prognosis, although probably not a clinically useful marker because serum levels are quite variable in normal humans. By immunohistochemistry it is a good marker. Barry Kacinski started this work with ovarian cancer where it is a good marker. For breast cancer there have been four independent studies with 100 patients or so in each. What about lung and prostate cancer? Only a couple of people were working on this until recently. People have now looked in prostate, and the rumour is that it is a good marker there. My hepatology friends tell me that liver cancers are highly populated with macrophages, and this might be a good one to look at. In a recent paper published in *Cancer Research* researchers took transplantable human tumours and used antisense to mouse CSF-1 given intravenously (Aharinejad et al 2002). The rate of growth, progression and metastatic capability of these tumours was reduced. One of the significant reductions was in angiogenesis. Since the anti-CSF-1 was directed against mouse, it wasn't affecting something made by the human. This could be a therapeutic option.

Gordon: I'd like to ask about the CSF-independent macrophages that are found not only in lymphoid organs, but also thymus and microglia. Is there a different growth factor that we haven't discovered yet?

Pollard: This is one of the things we have been thinking of. To my knowledge no one has looked at this.

Gordon: Do we have any insight into the specialized functions of those CSF-independent macrophages?

Pollard: The density of the CSF-1-independent macrophages and microglia are only very slightly affected by the absence of the CSF-1, if at all. We do know that there are defects in those macrophages in the absence of CSF-1. We have done quite a few recent experiments with *Listeria*. Splenic macrophages have normal densities in the *op/op* mouse, but recruitment of new macrophages and the ability to cope with *Listeria* is compromised. Those resident macrophages don't make things like IL10 and IL6. The consequence of this is that IFNγ is very much reduced. These macrophages therefore require CSF-1 for at least some functions (Guleria & Pollard 2001).

Harris: Now you have such an elegant model for pulling the macrophages in and activating them, you also have a model for trying to replace individual factors that the macrophages secrete. Can you make them secrete signalling molecules, to see if you can reproduce the effect of the progression of tumour or the metastasis?

Pollard: What we are really trying to set up is a reproducible model in a transplant situation so we can make the macrophages, which we reintroduce, different in terms of what they express. We must remember that although macrophages are the targets, if you look at the infiltrate it is not just macrophages that are coming in. Macrophages are therefore sending out signals to other cells. The final effector could be a macrophage product but it could also be a mast cell product or a neutrophil.

Gordon: I was struck by how few of those cells were macrophages. Are the others mostly tumour cells?

Pollard: They are mostly other blood cells, such as neutrophils or mast cells. There is a lot of fibrous material around those infiltrates. A lot of collagen is being laid down. If you do immunostain for F4/80 you will see that about half the cells are macrophages.

References

Aharinejad S, Abraham D, Paulus P et al 2002 Colony-stimulating factor-1 antisense treatment suppresses growth of human tumor xenografts in mice. Cancer Res 62:5317–5324
Guleria I, Pollard JW 2001 Aberrant macrophage and neutrophil population dynamics and impaired Th1 response to *Listeria monocytogenes* in colony-stimulating factor-1 deficient mice. Infect Immun 69:1795–1807

Chemokines: angiogenesis and metastases in lung cancer

Robert M. Strieter, John A. Belperio, Roderick J. Phillips and Michael P. Keane

Department of Medicine, Division of Pulmonary and Critical Care Medicine, University of California, Los Angeles, David Geffen School of Medicine, UCLA, 900 Veteran Avenue, 14–154 Warren Hall, Box 711922, Los Angeles, CA 90024

Abstract. Non-small cell lung cancer (NSCLC) growth, angiogenesis, invasion, and metastases to specific organs is dependent on an orchestrated series of events that include: cellular transformation; establishment of a pro-angiogenic environment; tumour cell proliferation, invasion and entry into the circulation; and tumour cell trafficking and metastatic tumour growth in specific organs based on the concept of Paget's theory of 'seed and soil' related to homing of tumour cells to a specific organ. The events that destine a tumour cell to invade and metastasize to distant organs are analogous to leukocyte trafficking. Chemokines have had an increasingly important role in mediating the homing of leukocytes under both conditions of homeostasis and inflammatory/immunological responses. Recently, the biology of chemokines has extended beyond their role in mediating leukocyte trafficking. Specifically, CXC chemokines have been found to be important in the regulation of angiogenesis, and in promoting tumour cell migration and organ-specific metastases. Data will be presented to highlight the importance of CXC chemokine ligands and receptors in mediating NSCLC tumour-associated angiogenesis, 'immunoangiostasis', and organ specific metastases. These findings may ultimately lead to clinical strategies to attenuate the pathobiology of CXC chemokines in promoting NSCLC tumour growth and metastases.

2004 Cancer and inflammation. Wiley, Chichester (Novartis Foundation Symposium 256) p 173–188

Non-small cell lung cancer (NSCLC) tumour growth, invasion, and metastasis to specific organs is dependent on a highly orchestrated series of events that include: cellular transformation; establishment of a pro-angiogenic environment; local tumour cell proliferation; local invasion through extracellular matrix (ECM)/ vascular basement membrane and entry into the circulation; and metastatic tumour cell trafficking and growth in specific organs. The above events that destine a tumour cell to invade and metastasize to specific organs are analogous to leukocyte entry into the circulation, and eventual homing to specific tissue sites. Chemokines play a critical role in mediating the homing of leukocytes under both homeostasis and inflammatory/immunological responses. Over the last decade studies have demonstrated that chemokines/chemokine receptors can directly promote tumour-associated angiogenesis, tumour growth and metastases. We will review the role that chemokines play in mediating angiogenesis and metastases of lung cancer.

Chemokines are involved in the regulation of angiogenesis

The neoplastic transformation, growth, survival, invasion, and metastases are dependent on the establishment of a pro-angiogenic environment. Local angiogenesis is determined by an imbalance in the overexpression of pro-angiogenic signals, as compared to inhibitors of angiogenesis. Members of the CXC chemokine family play disparate roles in the regulation of angiogenesis. CXC chemokines are heparin-binding proteins. They have four highly conserved cysteine amino acid residues, with the first two cysteines separated by one non-conserved amino acid residue, hence the name CXC (Balkwill 1998, Luster 1998, Rollins 1997). Although the CXC motif distinguishes this family from other chemokine families, a second structural domain within this family also dictates their biological activity. The N-terminus of some CXC chemokines contains three amino acid residues (Glu-Leu-Arg; 'ELR' motif) which precedes the first cysteine amino acid residue of the primary structure of these cytokines (Balkwill 1998, Luster 1998, Rollins 1997). Members that contain the 'ELR' motif (ELR$^+$) are potent promoters of angiogenesis in physiological concentrations of 1–100 nM (Strieter et al 1995). In contrast, members that lack the ELR motif (ELR$^-$) and are interferon-inducible are potent inhibitors of angiogenesis in physiological concentrations of 500 pM to 100 nM (Strieter et al 1995). On a structural/ functional level the CXC chemokine family behaves in a disparate manner in the promotion or inhibition of angiogenesis relevant to NSCLC.

Angiogenic (ELR$^+$) CXC chemokines

The angiogenic members of the CXC chemokine family include interleukin 8 (IL8/CXCL8), epithelial neutrophil activating protein 78 (ENA78/CXCL5), growth related genes (GROα/CXCL1 -β/CXCL2 and -γ/CXCL3), granulocyte chemotactic protein 2 (GCP2/CXCL6) and N-terminal truncated forms of platelet basic protein (PBP), which include connective tissue activating protein III (CTAP-III), β thromboglobulin (β-TG), and neutrophil activating protein 2 (NAP2/CXCL7) (Strieter et al 1995) (Table 1).

While angiogenic factors can behave in a redundant manner, recent studies have shown that serial interactions may ultimately contribute to perpetuation of the angiogenic phenotype of the endothelium. Nor and associates have reported that vascular endothelial cell growth factor (VEGF)→↑Bcl-2 and

TABLE 1   CXC Chemokine Family

- ELR$^+$ CXC chemokines are angiogenic factors
    - IL8 (CXCL8)
    - ENA78 (CXCL5)
    - GROα (CXCL1)
    - GROβ (CXCL2)
    - GROγ (CXCL3)
    - GCP2 (CXCL6)
        - PBP
        - CTAP-III
        - β-TG
        - NAP-2 (CXCL7)

- ELR$^-$ CXC chemokines are angiostatic factors
    - PF-4 (CXCL4)
- IFN-inducible ELR$^-$ CXC chemokines
    - IP10 (CXCL10)
    - MIG (CXCL9)
    - ITAC (CXCL11)

subsequent up-regulation/expression of IL-8/CXCL8 that acts in an autocrine and paracrine manner to maintain the angiogenic phenotype of the endothelial cells (Nor et al 2001). Human microvascular endothelial cells over-expressing Bcl-2 (HMVEC-Bcl-2), as well as VEGF-treated HMVECs exhibit a 15-fold and fourfold increase, respectively, in the expression of the pro-angiogenic CXC chemokine, IL8/CXCL8 (Nor et al 2001). Transfection of antisense Bcl-2 into HMVECs can block VEGF-mediated induction of IL8/CXCL8 (Nor et al 2001). Conditioned media from HMVEC-Bcl-2 induces angiogenic activity, which is, in part, dependent on the expression of IL8/CXCL8 (Nor et al 2001).

This pathway *in vitro* can be reproduced *in vivo* when human tumour cells that normally do not form tumours *in vivo* in SCID are mixed with human endothelial cells over-expressing Bcl-2 (HMVEC-Bcl-2). The combination of these cells generate human tumour growth with vascularized human microvessels (Nor et al 2001). Coimplantation of HMVEC-Bcl-2 and tumour cells resulted in enhanced tumour growth related to increased intratumoral microvascular density and enhanced endothelial cell survival (Nor et al 2001). Moreover, when SCID mice were implanted with tumour cells that produce VEGF and treated with neutralizing anti-IL8/CXCL8 antibodies, tumours exhibited a marked reduction in microvessel density and reduction in tumour volume (Nor et al 2001). These results demonstrate the existence of a novel serial pathway of VEGF→↑Bcl-2 in endothelial cells that leads to enhanced intratumoural microvascular survival/ density and tumour growth that is IL8/CXCL8-dependent.

CXCR2 is the putative receptor for angiogenic (ELR$^+$) CXC chemokine-mediated angiogenesis

The fact that all ELR$^+$ CXC chemokines mediate angiogenesis highlights the importance of identifying a common receptor. The candidate CXC chemokine receptors are CXCR1 and/or CXCR2. Only IL8/CXCL8 and GCP2/CXCL6 specifically bind to CXCR1, whereas, all ELR$^+$ CXC chemokines bind to CXCR2 (Balkwill 1998, Luster 1998, Rollins 1997). The ability of all ELR$^+$ CXC chemokine ligands to bind to CXCR2 supports the notion that this receptor mediates the angiogenic activity of ELR$^+$ CXC chemokines.

Addison and colleagues have demonstrated that CXCR2 is detected in HMVECs at both the mRNA and protein levels. In addition, the expression of CXCR2, not CXCR1, was found to be functional in mediating endothelial cell chemotaxis. Moreover, this response was sensitive to Pertussis toxin, suggesting a role for G protein-linked receptor mechanisms in mediating HMVEC chemotaxis (Addison et al 2000a). Furthermore, the importance of CXCR2 in mediating ELR$^+$ CXC chemokine-induced angiogenesis was demonstrated *in vivo* using the cornea micropocket assay of angiogenesis in CXCR2$^{+/+}$ and $^{-/-}$ animals. ELR$^+$ CXC chemokine-mediated angiogenesis was inhibited in the corneas of CXCR2$^{-/-}$ mice, and in the presence of neutralizing antibodies to CXCR2 in the rat corneal micropocket assay. These studies have now been extended to a syngeneic tumour model system in CXCR2$^{-/-}$, as compared to CXCR2$^{+/+}$ mice. Lung cancer in CXCR2$^{-/-}$ demonstrates reduced growth, increased tumour-associated necrosis, inhibited tumour-associated angiogenesis and metastatic potential. These *in vitro* and *in vivo* studies establish that CXCR2 is an important receptor that mediates ELR$^+$ CXC chemokine-dependent angiogenic activity.

Interferon (IFN)-inducible (ELR$^-$) CXC chemokines are inhibitors of angiogenesis

The angiostatic members of the CXC chemokine family include PF4, monokine induced by IFNγ (MIG/CXCL9), and IFNγ-inducible protein (IP10/CXCL10) (Balkwill 1998, Luster 1998, Rollins 1997) (Table 1). IP10/CXCL10 can be induced by all three interferons (IFNα, β and γ) (Balkwill 1998, Luster 1998, Rollins 1997). MIG/CXCL9 and IFN-inducible T-cell α chemoattractant (I-TAC/CXCL11), other interferon-inducible ELR$^-$ members of the CXC chemokine family, similar to IP10/CXCL10 and MIG/CXCL9, inhibits

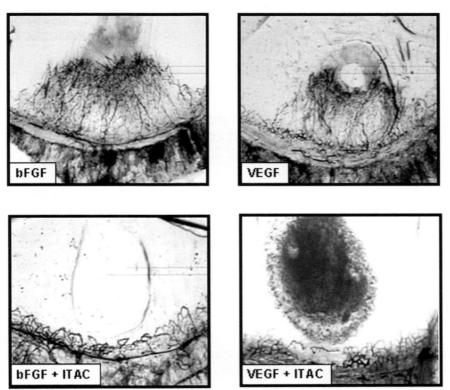

FIG. 1. The interferon-inducible ELR⁻ CXC chemokine, I-TAC/CXCL11, inhibits the angiogenic activity of bFGF, VEGF and IL8.

neovascularization in the CMP assay in response to either ELR⁺ CXC chemokines or VEGF (Fig. 1). These findings suggest that all IFN-inducible ELR⁻ CXC chemokines are potent inhibitors of angiogenesis. Moreover, this interrelationship of interferon and IFN-inducible CXC chemokines and their biological function are directly relevant to the function of other cytokines, such as IL18 and IL12 that stimulate the expression of IFNs. Therefore, cytokines such as IL12 and IL18, via the induction of IFNγ, will have a profound effect on the production of IP10/CXCL10, MIG/CXCL9, and I-TAC/CXCL11. The subsequent expression of IFN-inducible CXC chemokines may represent the final common pathway and explain the mechanism for the attenuation of angiogenesis related to interferons. Furthermore, this cytokine cascade interconnects with Th1-mediated immunity toward tumour-associated antigens and creates the concept of 'immunoangiostasis' (Fig. 2).

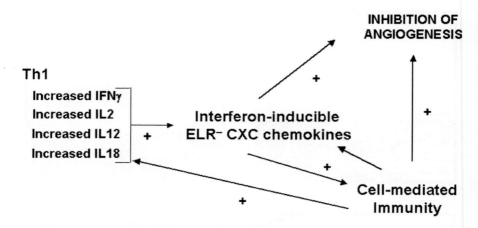

FIG. 2. The interrelationship of Th1 cytokine/cell-mediated immunity and interferon-inducible ELR⁻ CXC chemokines: concept of immunoangiostasis. The biology of interferon-inducible ELR⁻ CXC chemokines in the inhibition of angiogenesis is linked to cell-mediated immunity. This highlights the importance of an appropriate immune response to tumour-associated antigens, and the subsequent generation of angiostatic factors that can further modify local tumour-associated angiogenesis leading to the concept of immunoangiostasis.

CXCR3 is the putative receptor for angiostatic interferon-inducible (ELR⁻) CXC chemokine inhibition of angiogenesis

All three IFN-inducible ELR⁻ CXC chemokines specifically bind to the CXC chemokine receptor, CXCR3 (Balkwill 1998, Luster 1998, Rollins 1997). The original observation that CXCR3 was found on endothelium was shown in murine endothelial cells (Soto et al 1998). Romagnani and colleagues have identified that CXCR3 is expressed on human endothelium (Romagnani et al 2001). They showed that CXCR3 is expressed by a small percentage of HMVEC in several human normal and pathological tissues, including endothelial cells within tumours (Romagnani et al 2001). Cultures of HMVECs also express CXCR3, although this expression was limited to the S/G2-M phase of their cell cycle. Angiostatic IFN-inducible ELR⁻ CXC chemokines, IP10/CXCL10, MIG/CXCL9, and I-TAC/CXCL11 all blocked HMVEC proliferation *in vitro*. However, this effect was inhibited by an anti-CXCR3 antibody that inhibits ligand binding to the receptor (Romagnani et al 2001). These data provide definitive evidence of CXCR3 expression by HMVECs. Salcedo and associates have substantiated these findings for the expression of CXCR3 on endothelial cells and determined that IP10/CXCL10, MIG/CXCL9, and I-TAC/CXCL11 could inhibit the endothelial cell chemotactic response to IL8/CXCL8 (Salcedo et al 2000). These findings open new avenues for consideration of therapeutic interventions in the treatment of aberrant angiogenesis associated with lung cancer.

ELR⁺ CXC chemokines promote angiogenesis associated with tumorigenesis

The ELR⁺ CXC chemokines are important mediators of tumorigenesis related to their angiogenic properties. IL8/CXCL8 is markedly elevated and contributes to the overall angiogenic activity of NSCLC (Smith et al 1994). Extending these studies to an *in vivo* model system of human tumorigenesis (i.e. human NSCLC/SCID mouse chimera) (Arenberg et al 1996a), tumour-derived IL8/CXCL8 was found to be directly correlated with tumorigenesis (Arenberg et al 1996a). Tumour-bearing animals depleted of IL8/CXCL8 demonstrated a marked reduction in tumour growth and a reduction in spontaneous metastases (Arenberg et al 1996a). The attenuation of tumour growth and metastases was directly correlated to reduced angiogenesis. These findings have been further corroborated using several human NSCLC cell lines grown in nude mice. NSCLC cell lines that constitutively express IL8/CXCL8 display greater tumorigenicity that is directly correlated to angiogenesis (Yatsunami et al 1997).

While IL8/CXCL8 was the first angiogenic CXC chemokine to be discovered in NSCLC, ENA78/CXCL5 has now been determined to have a higher degree of correlation with NSCLC-derived angiogenesis (Arenberg et al 1998). Surgical specimens of NSCLC tumours demonstrate a direct correlation of ENA78/CXCL5 with tumour angiogenesis. These studies were extended to a SCID mouse model of human NSCLC tumorigenesis. ENA78/CXCL5 expression was directly correlated with tumour growth. Moreover, when NSCLC tumour bearing animals were depleted of ENA78/CXCL5, both tumour growth and spontaneous metastases were markedly attenuated (Arenberg et al 1998). The reduction of angiogenesis is also accompanied by an increase in tumour cell apoptosis, consistent with the previous observation that inhibition of tumour-derived angiogenesis is associated with increased tumour cell apoptosis (O'Reilly et al 1997). Similarly, *in vivo* and *in vitro* proliferation of NSCLC cells was unaffected by the presence of ENA78/CXCL5. While a significant correlation of ENA78/CXCL5 exists with tumour-derived angiogenesis, tumour growth, and metastases, ENA78/CXCL5 depletion does not completely inhibit tumour growth. This reflects that the angiogenic activity of NSCLC tumours is related to many overlapping and potentially redundant factors acting in a parallel or serial manner.

Interferon-inducible (ELR⁻) CXC chemokines
attenuate angiogenesis associated with tumorigenesis

Levels of IP10/CXCL10 from human surgical NSCLC tumour specimens have been found to be significantly higher in specific tumour specimens (Arenberg et al 1996b). IP10/CXCL10 levels are higher in squamous cell carcinoma (SCCA), as compared to adenocarcinoma. Moreover, depletion of IP10/CXCL10 from SCCA surgical specimens resulted in augmented angiogenic activity (Arenberg et al 1996b). The marked difference in the levels and bioactivity of IP10/CXCL10 in SCCA and adenocarcinoma is clinically and pathophysiologically relevant, and represents a possible mechanism for the biological differences between these two cell-types of NSCLC. Patient survival is lower, metastatic potential is higher, and evidence of angiogenesis is greater for adenocarcinoma, as compared to SCCA of the lung (Minna 1991). These studies were extended to a SCID mouse system to examine the effect of IP10/CXCL10 on human NSCLC cell line tumour growth in a T and B cell-independent manner. Mice were inoculated with either adenocarcinoma or SCCA cell lines (Arenberg et al 1996b). The production of IP10/CXCL10 from adenocarcinoma and SCCA tumours was inversely correlated with tumour growth (Arenberg et al 1996b). However, IP10/CXCL10 levels were significantly higher in the SCCA, as compared to adenocarcinoma tumours. The appearance of spontaneous lung metastases in mice bearing adenocarcinoma tumours occurred after IP10/CXCL10 levels from either the primary tumour or plasma had reached a nadir. In subsequent experiments, mice bearing SCCA tumours were treated with either neutralizing anti-IP10/CXCL10 antibodies, whereas, animals bearing adenocarcinoma tumours were treated with intra-tumoural IP10/CXCL10. Depletion of IP10/CXCL10 in SCCA tumours resulted in a twofold increase in their size. In contrast, reconstitution of intra-tumoural IP10/CXCL10 in adenocarcinoma tumours reduced both their size and metastatic potential; that was unrelated to infiltrating neutrophils or mononuclear cells and directly attributable to a reduction in tumour-associated angiogenesis.

Similar to IP10/CXCL10, the IFN-inducible ELR⁻ CXC chemokine, MIG/CXCL9, also plays a significant role in regulating angiogenesis of NSCLC. Addison and associates have found that MIG/CXCL9 levels in human specimens of NSCLC was not significantly different from that found in normal lung tissue (Addison et al 2000b). However, these results suggested that the increased expression of ELR⁺ CXC chemokines found in these tumour samples was not counter-regulated by a concomitant increase in the expression of the angiostatic interferon-inducible ELR⁻ CXC chemokine MIG/CXCL9. This imbalance would promote neovascularization. To alter this imbalance, they demonstrated that over-expression of the interferon-inducible ELR⁻ CXC chemokine, MIG/CXCL9, by three different strategies including gene transfer, resulted in the

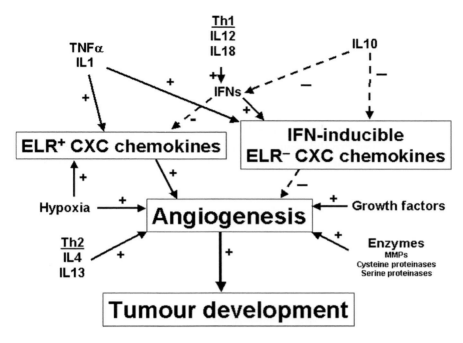

FIG. 3. The interrelationship of CXC chemokines with other factors in the regulation of angiogenesis that impacts on tumour development and metastatic potential.

inhibition of NSCLC tumour growth and metastasis via a decrease in tumour-derived vessel density. These findings support the importance of the interferon-inducible ELR⁻ CXC chemokines in inhibiting NSCLC tumour growth by attenuation of tumour-derived angiogenesis. The above studies demonstrate the potential efficacy of gene therapy as an alternative means to deliver and over-express a potent angiostatic CXC chemokine. Furthermore, these findings demonstrate the importance of CXC chemokines in the regulation of angiogenesis (Fig. 3).

Evidence that chemokines regulate the pattern of organ-specific metastasis of lung cancer

Paget in 1889 was the first to demonstrate that carcinoma has a distinct metastatic pattern preferentially involving the regional lymph nodes, bone marrow, lung, and liver (Paget 1889). Recently, Müller and colleagues (Müller et al 2001) have provided new insights into potential mechanisms related to organ-specific metastasis of breast cancer cells directly related to a CXC chemokine and its receptor. They found that CXCR4 was the most highly expressed chemokine receptor in human breast cancer. The ligand for CXCR4, SDF1/CXCL12 exhibited peak mRNA levels in organs that are preferential destinations of breast cancer metastasis. Moreover, *in vivo* neutralization of SDF1/CXCL12/CXCR4 interactions with specific anti-CXCR4 antibodies resulted in significant inhibition of metastasis of breast cancer in an organ-specific manner.

Although the above study has provided evidence that the predominant function of SDF1/CXCL12/CXCR4 in tumorigenesis is one of inducing metastasis, studies have also indicated that this chemokine receptor/ligand pair are potent angiogenic factors (Kijowski et al 2001, Salcedo et al 1999). Evidence suggests that SDF1/CXCL12α may be involved in up-regulating levels of vascular endothelial growth factor (VEGF) and basic fibroblast growth factor (bFGF) and that subcutaneous injection of SDF1/CXCL12α into mice induces formation of local small blood vessels (Kijowski et al 2001, Salcedo et al 1999). However, it has yet to be demonstrated in an *in vivo* tumour model system that endogenous SDF1/CXCL12 binding to CXCR4 mediates a significant portion of primary tumour angiogenesis, and angiogenesis-dependent tumour growth.

To address the role of the CXCR4/SDF1/CXCL12 axis in NSCLC, we developed both a tissue culture-based model system and an *in vivo* mouse paradigm to characterize the activity of this receptor/ligand pair. Our results indicate that NSCLC cells express CXCR4, but not its ligand SDF1/CXCL12. SDF1/CXCL12 stimulation of CXCR4 on NSCLC cells leads to chemotaxis, calcium mobilization, and activation of mitogen-activated protein kinase p42/44. In addition, SDF1/CXCL12 protein levels are significantly higher in those organs that are known to be highly susceptible for NSCLC metastasis, as compared to the primary tumour and plasma levels, suggesting that a chemotactic gradient could be established between the site of the primary tumour and those organs that develop NSCLC tumour metastases. To determine whether CXCR4 expressing tumour cells have a selective advantage for metastases, primary tumour and metastases were assessed for expression of CXCR4. Sixty five percent of the tumour cells in the primary tumour were found to express CXCR4, suggesting that 35% are actually CXCR4 negative. However, when the metastases were isolated from the adrenal glands, liver, lungs and bone marrow of tumour-bearing SCID mice, greater than 99% of the NSCLC tumour cells were expressing CXCR4. Moreover, analysis of the heart and kidney revealed few if any metastases. Therefore, it seems likely that, in common with breast cancer cells, CXCR4 plays an important role in NSCLC tumour metastasis. Furthermore, *in vivo* neutralization of SDF1/CXCL12 in a SCID mouse system of spontaneous metastasis of human NSCLC resulted in marked attenuation of NSCLC metastases to several organs including the adrenal glands, liver, lung and bone marrow. At the same time, SDF1/CXCL12/CXCR4 does not promote tumour-associated angiogenesis or tumour growth *in vivo* in primary tumours. Thus these findings provide strong evidence to suggest that the SDF1/CXCL12/CXCR4 axis may be important in orchestrating the metastasis of NSCLC cells to a number of organs throughout the host in an angiogenesis-independent manner.

Conclusion

Although chemokine biology was originally felt to be restricted only to recruitment of populations of leukocytes, it has become increasingly clear that these cytokines can display pleiotropic effects in mediating biology that goes beyond their originally described function. There is no better human disease to study diversity of chemokine function than tumour biology. Chemokines have an autocrine, paracrine, and hormonal role at every level related to primary tumour growth, invasion, and metastasis to distant preferential organs. The understanding of this expanded role of chemokines in tumour biology will provide novel opportunities for therapeutic intervention.

Acknowledgments

This work was supported, in part, by: NIH grants HL66027, CA87879, P50 CA90388, and P01 HL67665 (RMS), HL04493 (JAB) and HL03906 and P01 HL67665 (MPK).

References

Addison CL, Daniel TO, Burdick MD et al 2000a The CXC chemokine receptor 2, CXCR2, is the putative receptor for ELR$^+$ CXC chemokine-induced angiogenic activity. J Immunol 165:5269–5277

Addison CL, Arenberg DA, Morris SB et al 2000b The CXC chemokine, monokine induced by interferon-gamma, inhibits non-small cell lung carcinoma tumor growth and metastasis. Hum Gene Ther 11:247–261

Arenberg DA, Kunkel SL, Polverini PJ, Glass M, Burdick MD, Strieter RM 1996a Inhibition of interleukin-8 reduces tumorigenesis of human non-small cell lung cancer in SCID mice. J Clin Invest 97:2792–2802

Arenberg DA, Kunkel SL, Polverini PJ et al 1996b Interferon-gamma-inducible protein 10 (IP-10) is an angiostatic factor that inhibits human non-small cell lung cancer (NSCLC) tumorigenesis and spontaneous metastases. J Exp Med 184:981–992

Arenberg DA, Keane MP, DiGiovine B et al 1998 Epithelial-neutrophil activating peptide (ENA-78) is an important angiogenic factor in non-small cell lung cancer. J Clin Invest 102:465–472

Balkwill F 1998 The molecular and cellular biology of the chemokines. J Viral Hepat 5:1–14

Kijowski J, Baj-Krzyworzeka M, Majka M et al 2001 The SDF-1-CXCR4 axis stimulates VEGF secretion and activates integrins but does not affect proliferation and survival in lymphohematopoietic cells. Stem Cells 19:453–466

Luster AD 1998 Chemokines—chemotactic cytokines that mediate inflammation. N Engl J Med 338:436–445

Minna JD 1991 Neoplasms of the lung. In Isselbacher KJ (ed) Principles of internal medicine. 12th edn, New York, McGraw-Hill, p 1102–1110

Müller A, Homey B, Soto H et al 2001 Involvement of chemokine receptors in breast cancer metastasis. Nature 410:50–56

Nor JE, Christensen J, Liu J et al 2001 Up-regulation of Bcl-2 in microvascular endothelial cells enhances intratumoral angiogenesis and accelerates tumor growth. Cancer Res 61:2183–2188

O'Reilly MS, Boehm T, Shing Y et al 1997 Endostatin: an endogenous inhibitor of angiogenesis and tumor growth. Cell 88:277–285

Paget SL 1889 The distribution of secondary growths in cancer of the breast. Lancet 1:571

Rollins BJ 1997 Chemokines. Blood 90:909–928

Romagnani P, Annunziato F, Lasagni L et al 2001 Cell cycle-dependent expression of CXC chemokine receptor 3 by endothelial cells mediates angiostatic activity. J Clin Invest 107:53–63

Salcedo R, Wasserman K, Young HA et al 1999 Vascular endothelial growth factor and basic fibroblast growth factor induce expression of CXCR4 on human endothelial cells: in vivo neovascularization induced by stromal-derived factor-1alpha. Am J Pathol 154:1125–1135

Salcedo R, Resau JH, Halverson D et al 2000 Differential expression and responsiveness of chemokine receptors (CXCR1-3) by human microvascular endothelial cells and umbilical vein endothelial cells. FASEB J 14:2055–2064

Smith DR, Polverini PJ, Kunkel SL et al 1994 Inhibition of IL-8 attenuates angiogenesis in bronchogenic carcinoma. J Exp Med 179:1409–1415

Soto H, Wang W, Strieter RM et al 1998 The CC chemokine 6Ckine binds the CXC chemokine receptor CXCR3. Proc Natl Acad Sci USA 95:8205–8210

Strieter RM, Polverini PJ, Kunkel SL et al 1995 The functional role of the ELR motif in CXC chemokine-mediated angiogenesis. J Biol Chem 270:27348–27357

Yatsunami J, Tsuruta N, Ogata K et al 1997 Interleukin-8 participates in angiogenesis in non-small cell, but not small cell carcinoma of the lung. Cancer Lett 120:101–108

DISCUSSION

Oppenheim: I couldn't follow your explanation of the interaction between SLC or CCL21 with the IP10. Could you explain this again?

Strieter: If we take CCL21, it has a dramatic effect in promoting tumour reduction. This appears to be a combination of the recruitment of specific subsets of cells. It is a CD4/CD8-dependent as well as perhaps a CD4/CD8-independent system. At the same time there is also a role for CCL21 in the mouse system to actually inhibit angiogenesis in a direct manner. It is also doing this through IFNγ, so it is IFNγ dependent. If we block this we see a significant attenuation of the ability of CCL21 to inhibit tumour growth. When we look further downstream in terms of finding a common final pathway leading to the interferon induction of chemokines, we also found a marked increase in both MIG/CXCL9 and IP10/CXCL10. Then when we attenuate the expression of these molecules we see that they have a significant impact similar to what we see with IFNγ in terms of cellular recruitment as well as in terms of altering their ability to promote angiostasis.

Oppenheim: So the CCL21, by promoting Th1-type IFNγ-dependent responses has anti-angiogenic and therefore antitumour effects which may be mediated and amplified by the downstream induction of ligands (e.g. IP10 and MIG for CXCR3).

Rollins: I want to confirm and expand a little bit on the expression of CXCR4 by NSCLC cell lines. We published a study (Soejima & Rollins 2001) looking at about 10 or 11 of these lines. We saw CXCR4 in all of them. However, we also saw CXCR4 on small cell lung cancer cell lines. We know that CXCR4 is essentially on all breast cancer cell lines, all NSCLC cell lines and all small cell lung cancer cell lines. It is important to take things back to the human situation. Even though the organs involved by metastatic disease in all three of these primary carcinomas overlap somewhat, the proportions are not the same for each one. If we are in the position of wanting to ascribe metastatic spread to the expression of CXCR4 and SDF1, that clearly cannot be the whole story. There has to be something else to explain what we see in the clinic: these sorts of differential patterns of spread. We have to keep this kind of thing in mind and try to think about what the other factors are.

Strieter: The data that the depletion of SDF1/CXCL12 does not completely attenuate all the metastatic depletion, but that essentially all of the metastatic cells express CXCR4 would support your contention that CXCR4 is a cofactor and other factors may be involved. I agree that there must be other factors.

Richmond: Barrett Rollins, are you saying that the organ-specific metastasis sites are different for each of those tumours, even though they all express CXCR4?

Rollins: I am saying that clinically the proportion of sites that are involved differ among them. In small cell lung cancer, for example, we see much more bone marrow and CNS involvement than in breast cancer, although both things happen in breast cancer as well.

Richmond: So it is possible that arriving isn't the only thing that you are looking at, but the ability of the cells to proliferate after they arrive is also important here.

Rollins: That would be another explanation.

Balkwill: The CXCR4 issue is bigger than just breast cancer, non-small cell and small cell lung cancer. I will describe our data in ovarian cancer data in my paper (Szlosarek & Balkwill 2004, this volume). Cancer cells in glioblastoma, neuroblastoma, pancreatic and prostate cancer all express CXCR4. Why is CXCR4 found on the majority of tumour cells? Is CXCR4 expression tumour specific? For instance we don't find CXCR4 on normal ovarian surface epithelium. Is this true also for the NSCLCs?

Strieter: There is constitutive expression of CXCR4 on the lung, but it is not at the same magnitude as in the tumour.

Rollins: In our report we looked at normal human bronchial epithelial cells and compared them to these non-small cell lines. We saw CXCR4 expressed at the same level. The caveat there is that we are looking at cell lines.

Balkwill: Is it perhaps a difference in signalling in the tumour cells compared with the normal cells? Is that what we should be looking at?

Richmond: Yes, that is entirely possible. There is inherent activation of many of the molecules that are involved in the signalling pathway presensitizing those tumour cells to a signal and amplifying it considerably. It could also go the other way.

Balkwill: It is still a question of why cancer cells express CXCR4. Is it because CXCR4 is found in stem cells normally?

Strieter: Another way to look at this is to say that it has usurped what is normally a homeostatic trafficking system, and that there is evidence that the ligand is expressed in a greater fashion and we are seeing organs that are obviously predisposed to these metastatic burdens. It is a system which the tumour cell has basically usurped from normal leukocyte trafficking. Why CXCR4 as opposed to other chemokines again may be based on the fact that we are dealing with homeostatic as opposed to inflammatory induction of the receptor.

Oppenheim: SDF stands for 'stromal-derived factor'. Wherever there is stroma, which is almost everywhere, there is the potential for producing SDF. The sites which are particular targets for metastatic spread are those rich in stroma. The other thing that is unique for CXCR4 and SDF1 is that they are more developmentally involved than some of the other chemokines. Knocking out SDF has a very dramatic effect and such mice live only a few days if they are born at all. They have defects in vascularization, particularly in the mesentery and perhaps in the cerebellum, and there are defects in the heart. This is unlike the other chemokines, where there is apparent redundancy in angiogenesis accounting for the lack of defects in the knockouts. So the angiogenic role of CXCR4 is developmentally important as well. The idea that stem cells may be expressing it is an interesting one in terms of subsequent expression of tumour cells being rather common.

Gordon: Very few cancers spread to the spleen, even though there is a lot of stroma and all kinds of potentially attractive ligands. Why is this?

Strieter: I didn't specifically focus on the spleen in that regard. We focused on the organs in humans which are predisposed to metastases.

Harris: Human melanoma is the one that goes to the spleen. This is not uncommon.

Richmond: We have been having a problem in trying to coexpress some of these receptors in the same cell *in vitro*. *In vivo* there is a sort of yin–yang: if you have one of these receptors at a high level, you may not have the other. We find this to be true to an extent for CXCR2 and CXCR3. It is hard to get a cell line to co-express both of these receptors. Have you identified specific cells that express both of those receptors at the same time?

Strieter: The problem with that is the cell cycle story. We have had problems seeing both CXCR2 and CXCR3 on the endothelium at the same time. This goes back to the predominant potential expression of these during the cell cycle. We are trying to dissect out whether the expression pattern changes with the phase of cell cycle. This is an important issue that needs to be better defined.

Richmond: In nature it seems that cytokines that up-modulate CXCR3 down-modulate CXCR2, for example in T cell populations. Is there an innate mechanism that permits this, more than just the cell cycle? Or are there different populations of endothelial cells *in vivo* that are responding differently to those two ligands?

Strieter: I think that is probably the case.

Harris: Have you tried double blockade of CXCR2 and 3 in your models?

Strieter: No.

Harris: It would be interesting to know the expression patterns of the receptors in the blood vessels in the same tumour, to see if they were co-expressed.

Ristimäki: In both of the previous talks proteases, and in particular MMPs, have been brought up. Have you been able to test whether protease inhibitors work in your models? Do they inhibit metastasis?

Pollard: In the model we use this hasn't been tested.

Strieter: I haven't used them in my model.

Richmond: For colon cancer there is a group in Nebraska (Singh and colleagues), who have shown that IL8 induces MMP9. This facilitates the metastatic spread of these cells.

Pollard: MMP9 is the one that keeps popping up, although uPA is also involved.

Oppenheim: We did a study in which we demonstrated inhibition of tumour growth by anti-EGFR. Anti-IL8 alone did nothing, but together with anti-EGFR it had a synergistic antitumour effect. This was also capable of inducing MMP9.

Ristimäki: This is clearly a big field where academics are responsible because the medical firms have dropped the subject altogether. If there is something to it, it is us who need to work on this.

Pollard: One of the problems with proteases is that they are so overlapping that the phenotypes of the knockouts are not that severe until you start making double or triple knockouts.

Caux: I have a question regarding the CXCR2 ligands. You showed that they are produced by tumour cells themselves. Do you think they could also be produced by macrophages?

Strieter: They can be produced by all somatic cells depending on the signal. Fibroblasts, epithelial cells, tumour cells, macrophages and neutrophils can all produce these ELR$^+$ CXC chemokines. The macrophages are clearly a major cellular source of these in tumours. We showed that in squamous cell carcinoma the major cellular source was macrophages and fibroblasts, not tumour cells. Yet in adenocarcinoma the major source is the tumour cells. It varies.

Oppenheim: I know you limited your talk to CXC chemokines. There is evidence in the literature that there are a number of other chemokines in the CC and CX3C families, MCP-1, eotaxin, I309 and fractalkine, all of which have receptors expressed on microvascular-type endothelial cells. They also have proangiogenic effects. Presumably more papers will come out that may show that proinflammatory chemokines *per se* may have a lot of these effects, and that part and parcel of inflammation may be capacity to stimulate endothelial cells. It is necessary in inflammation to generate new blood vessels. Of course IL1 and TNF do this indirectly by generating chemokines and other angiogenic factors. Furthermore, VEGF cooperates with CXCR4 and SDF. Chemokines interact with other proinflammatory cytokines and VEGF in promoting angiogenesis.

References

Soejima K, Rollins BJ 2001 A functional IFN-gamma-inducible-protein-10/CXCL10-specific receptor expressed by epithelial and endothelial cells which is neither CXCR3 nor glycosaminoglycan. J Immunol 167:6576–6582
Szlosarek P, Balkwill F 2004 The inflammatory cytokine network of epithelial cancer: therapeutic implications. In: Cancer and inflammation. Wiley, Chichester (Novartis Found Symp 256), p 227–240

Macrophage infiltration and angiogenesis in human malignancy

Helen Knowles, Russell Leek* and Adrian L Harris[1]

*Cancer Research UK, Molecular Oncology Laboratories, Weatherall Institute of Molecular Medicine and *Nuffield Department of Clinical Laboratory Sciences, John Radcliffe Hospital, Oxford OX3 9DU, UK*

Abstract. It is well recognized that human tumours are hypoxic compared to normal adjacent tissues and that hypoxia is related to a poor outcome regardless of modality of treatment, including surgery alone, radiotherapy or chemotherapy. Hypoxia regulates a complex programme of gene transcription via hypoxia-inducible factors 1 and 2 (HIF-1, -2). We have shown that in breast cancer and many other tumour types, tumour-associated macrophages express high levels of HIF-2α compared to normal tissue macrophages and compared to the tumour. This high macrophage HIF-2α is an independent prognostic factor for poor outcome. The mechanisms up-regulating HIF-2α in macrophages may include inflammatory cytokines as well as hypoxia. Differentiation of monocytes into macrophages increases the basal level of HIF-2α protein and changes the programme of hypoxia. Many of these inducible genes are involved in inflammation and angiogenesis. Thus, the conversion of a peripheral monocyte into a macrophage generates a complex new programme of hypoxia-responsive genes that may contribute to angiogenesis and the complex microenvironment within the tumour, and as such provides important targets for therapy.

2004 Cancer and inflammation. Wiley, Chichester (Novartis Foundation Symposium 256) p 189–204

It is clear, examining tumour histology, that neoplastic cells exist in a complicated cellular milieu comprising the tumour cells themselves, stroma, vasculature and inflammatory cells. Indeed, cancers have been termed wounds that do not heal because of the contribution of angiogenesis and inflammatory cells to their behaviour. Amongst the stromal cells, the vascular endothelial cells have been extensively investigated in the last decade and the major contribution of angiogenesis towards tumour growth has been recognized. Angiogenesis is important for cellular growth, expansion of the tumour mass, migration,

---

[1]This paper was presented at the symposium by Adrian L. Harris to whom all correspondence should be addressed.

metastasis and the establishment of distant disease. It has become an important investigative target for anticancer therapy, with some encouraging early results with drugs that block the vascular endothelial growth factor receptor.

In spite of the increase in angiogenesis, it has been well recognized that human tumours are hypoxic compared to normal adjacent tissues with a pO_2 of around 1% oxygen (10 millimetres of mercury). Hypoxia itself is related to poor outcome with regards to the modality of treatment including surgery, radiotherapy or chemotherapy. One of the mechanisms is the up-regulation of a complex gene programme via the hypoxia-inducible transcription factors HIF-1 and HIF-2. Amongst the downstream genes are many angiogenic factors, particularly vascular endothelial growth factor.

Adding further complexity to the above interactions between hypoxia and angiogenesis are the role of inflammatory cells. Our work and that of others has shown that there is an association of macrophages with high angiogenesis in primary tumours (see below). Macrophages themselves produce many angiogenic factors as well as proteases and other cytokines that could help progression of cancer (Table 1). Macrophages are attracted towards hypoxic areas and clinical studies have shown that they tend to cluster away from the vessels in several tumour types. This review analyses the potential pathways by which macrophages can be attracted to tumours and switch on a hypoxia-regulated gene programme, relevant to progression of human malignancy.

From the evidence of these associations, it is possible to consider new therapeutic approaches involving targeting the macrophage or using the macrophage to deliver gene therapy. Although previous studies have suggested that macrophages may be important for anti-tumour effects in cancer, much of the evidence produced below will show that they are much more likely to be involved with progression and an aggressive phenotype in cancer.

Macrophage activities at tumour sites

Once a circulating monocyte has entered the tumour parenchyma it is referred to as a tumour associated macrophage (TAM) and it is capable of exhibiting a number of different behaviours depending upon its activation state. The nature of the activating signal in tumours is still unclear, but probably involves the interplay of factors expressed by the tumour cells and tumour micro-environmental conditions such as pH. Hypoxia seems to be involved in the process and it has been observed that the hypoxia inducible transcription factors HIF-1 and HIF-2 are both up-regulated in TAMs (Talks et al 2000). Essentially the activities of TAMs can be categorised into either anti- or pro-tumour effects. It had long been assumed that the major function of TAMs was as part of the host defence against developing tumours because macrophages were capable of killing tumour cells in a

TABLE 1   Angiogenic factors produced by macrophages

Factor	Abbreviation	Function
Vascular endothelial growth factor	VEGF	Angiogenic cytokine
Fibroblast growth factor 2, also known as basic FGF	FGF2	Angiogenic cytokine and ECM modulating cytokine
Epidermal growth factor	EGF	Angiogenic cytokine and tumour mitogen
Tumour necrosis factor alpha	TNFα	Angiogenic cytokine and ECM modulating cytokine
Thymidine phosphorylase	TP	Angiogenesis promoting enzyme
Hepatocyte growth factor/scatter factor	HGF/SF	Angiogenic cytokine
Insulin-like growth factor 1	IGF1	Angiogenic cytokine
Interleukin 8	IL8	Angiogenic cytokine
Collagenase		ECM degrading enzyme
Stromelysin		ECM degrading enzyme
Hyaluronidase	HA	ECM degrading enzyme
Urokinase-type plasminogen activator	u-PA	ECM degrading enzyme
Cathepsin D	Cath-D	ECM degrading enzyme
Tissue inhibitor of metalloprotease	TIMP	ECM enzyme inhibitor
Plasminogen activator inhibitor 1	PAI-1	ECM enzyme inhibitor
Transforming growth factor β	TGFβ	ECM modulating cytokine
Angiotropin		ECM modulating cytokine
Platelet-derived growth factor	PDGF	ECM modulating cytokine
Interleukin 6	IL6	ECM modulating cytokine

number of assays (Wilbanks et al 1999), but that this host reaction was eventually overwhelmed by the aggressive, fast growing tumour. Indeed this could well be true, and may account for the fact that there are more incidences of *in situ* cancers than invasive ones. It is also true that this host response is ineffective in many tumour types that eventually progress to an invasive and ultimately metastatic phenotype. Tumours may evade the immune response by altering HLA class 1 expression (Kaklamanis et al 1995) or by secreting factors such as interleukin (IL)4, IL6, IL10, transforming growth factor (TGF)β, PGE2, and colony stimulating factor 1 (CSF-1, also known as M-CSF) which down-regulate the immune functions of macrophages (Elgert el al 1998).

Recently it has become more apparent that tumours, in addition to merely evading the TAMs effects, may also be subverting the TAMs normal behaviours towards the tumours own advantage. Evidence for this comes from the fact that high levels of macrophage infiltration have been found to be associated with poor prognosis in many tumour types (Bingle et al 2002). Macrophages are potentially capable of promoting tumour progression through several different mechanisms. TAMs can produce a vast array of factors including mitogenic cytokines such as epidermal growth factor (EGF), the receptor for which is up-regulated in breast cancers of poor prognosis and is associated with macrophage infiltration, suggesting that EGF-producing TAMs may select for aggressive tumour phenotypes (Leek et al 2000). Importantly, TAMs also produce angiogenic factors like VEGF, TNFα and TP. They can also secrete proteolytic enzymes that digest the extracellular matrix releasing heparin bound angiogenic factors such as fibroblast growth factor (FGF)2 whilst simultaneously facilitating endothelial cell migration and reducing tumour cell adhesion, thus promoting metastasis (Leek et al 1997).

Tumour-promoting effects of TAMs and angiogenesis

TAMs are able to promote tumour progression through many mechanisms including direct mitogenic stimulation of neoplastic cells. However, one major pathway is via the stimulation of tumour angiogenesis (Leek et al 1997). It has been known for some time that macrophages express angiogenic cytokines and other angiogenic factors (Sunderkotter et al 1994). Initially no direct association was found between macrophage infiltration and angiogenesis in human neoplasias, although it had been observed that macrophage infiltration was associated with a poor prognosis in breast cancer (Steele et al 1984). The relationship between angiogenesis and prognosis was also unclear until Folkman and Weidner discovered that rather than an overall high degree of angiogenesis, only three 'hotspots' of high angiogenic activity were necessary to push a tumour into an unfavourable prognosis. This was because these 'hotspots' were biologically important as points of entry for metastatic tumour cells.

When TAM infiltration was examined in the same way in a series of breast cancers, it was noted that focal TAM infiltration had a major effect on prognosis. Tumours with macrophage indices (MØI) above the median had a much poorer relapse free and overall survival than those below the median, and patients with high MØI were more likely to die of their disease than those with low MØIs. When TAM infiltration was then compared to mean vessel density (MVD), it was apparent that a definite relationship existed between MØI and MVD as they were positively associated with each other ($P=0.03$) (Leek et al 1996). There was an association between macrophage infiltration and VEGF expression in tumour cells (Leek et al 2000). Given the wealth of evidence demonstrating the angiogenic capabilities of TAMs it seemed reasonable to surmise that the effect on prognosis was due, at least in part, to the induction of angiogenesis by the TAM population.

A later study demonstrated that the angiogenic factors thymidine phosphorylase and TNFα could be involved in the angiogenesis activities of macrophages. It has been demonstrated that TP can be up-regulated by TNFα in experimental assays. In breast tumours TNFα is produced predominantly by the TAM population and its expression in TAMs has been correlated with TP up-regulation in human breast carcinomas (Leek et al 1998) which may represent one axis of hypoxia independent TAM influence on angiogenesis. However, TAM clusters are spatially separated from vessel hotspots, and are found in the avascular areas between the vessel hotspots and necrotic zones (Leek et al 1996). The implication of this is that they are specifically attracted to these areas and hypoxia could play a role in switching of macrophages to an angiogenic phenotype. It is interesting to note that the most angiogenic tumours are also the most necrotic and have the highest macrophage indices indicating an important role for tumour hypoxia in the angiogenic activation of macrophages (Leek et al 1999).

The monotactic cytokine VEGF is hypoxically regulated via HIF-1 and may play a role in trafficking macrophages into these areas (Leek et al 2000). The nature of the angiogenic switch in TAMs is still poorly understood but it seems that hypoxia is a prime candidate for the stimulus. A recent study of TAMs in breast cancers has shown that they express both HIF-1 and HIF-2, and that HIF-2 expression is exclusive to the TAM population in these tumours (Talks et al 2000) (Figs 1, 2). This could mean that HIF-2 is the specific angiogenic switch activating macrophages into an angiogenic phenotype. When the numbers of HIF-2 expressing macrophage clusters were counted and compared to MVD in a recent study in a series of breast tumours, it was found that a very strong correlation existed between increased numbers of HIF-2 expressing macrophages and high levels of angiogenesis ($P=0.0001$) (Leek et al 2002). In this study it was also found that an inverse relationship existed between tumour TP expression and TAM HIF-2 ($P=0.0007$).

TP is generally not hypoxically regulated suggesting that two TAM-mediated angiogenic pathways may be in operation. Firstly, in hypoxic tumours, via the induction of macrophage HIF-2. Secondly, in the presence of oxygen, TP may induce tumour cell oxidative stress by the generation of reactive oxygen species which up-regulate the angiogenic factors VEGF, matrix metalloprotease 1 and IL8 (Brown et al 2000). These pathways seem to be reciprocal in that either one can be dominant dependent upon the oxygenation state of the tumour.

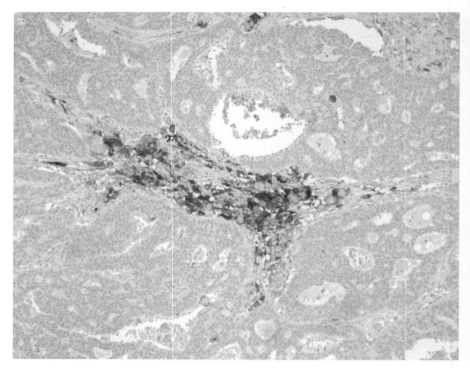

FIG. 1.   Low power view of a HIF-2α-positive macrophage cluster in an invasive breast tumour.

Tumour-associated macrophages and lymphangiogenesis

Whether lymphangiogenesis, i.e. the growth of new lymphatics, occurs in human tumours or not is controversial. We have recently shown in head and neck cancers that proliferating lymphatics are detected using a new antibody specific to lymphatics, LYVE-1 (Beasley et al 2002). Although this occurs in head and neck cancers, the opposite was the case in breast cancer where there was a peritumoural loss of lymphatics (Williams et al 2003). It appeared the tumours that had the most aggressive phenotype had the widest rim of lymphatic destruction and this also correlated with the macrophage infiltrate. The hypothesis was put forward that in the process of aggressive tumour cells invading the extracellular matrix around the tumour, the lymphatics were destroyed. Nevertheless this aggressive phenotype allowed the tumour cells to enter the lymphatics and spread to regional lymph nodes. The above two studies highlight the differences between tumour types in lymphatics and their regulation. It has recently been reported that yet another pattern appears in human cervical cancer (Schoppmann et al

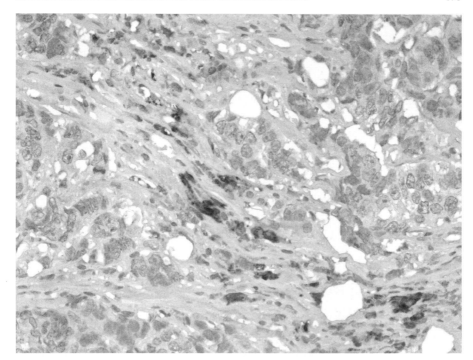

FIG. 2.   High power view of HIF-2α-positive macrophages an invasive breast tumour.

2002). In this study it was shown that there was a significant increase in lymphatic microvessels in peritumoural stroma and that the cells expressing most VEGF-C and VEGF-D, the main lymphatic growth factors, were stromal cells. The VEGF-C and VEGF-D-producing stromal cells were a subset of activated tumour associated macrophages which also expressed the receptor for VEGF-C and -D, VEGFR3. Peripheral blood monocytes were shown to express VEGFR3 but did not express VEGF-C and -D. After activation with lipopolysaccharide or tumour necrosis factor they were able to produce VEGF-C, suggesting that tumour associated macrophages may also play another role in peritumoural lymphoangiogenesis and dissemination in human cancer.

Studies with knockout mice and macrophages

The link between macrophage infiltration, expression of markers of angiogenesis and subsequent tumour progression has also been demonstrated in murine tumour models. Human tumour xenografts grown in matrix metalloproteinase 9 knockout

(MMP9$^{-/-}$) mice, which produce MMP9$^{-/-}$ macrophages, are smaller than those grown in MMP9$^{+/+}$ animals and have a reduced level of macrophage infiltration (Huang et al 2002). This is associated with a reduced microvessel density and decreased tumour VEGF concentration. MMP9 promotes the migration and invasion of cancer cells by mediating type IV collagen degradation in the vascular basement membrane. MMP9$^{-/-}$ macrophages were therefore less able to penetrate the extracellular matrix, resulting in the observed reduction in macrophage infiltration and tumour angiogenesis (Huang et al 2002). The importance of the matrix metalloproteinases in regulating macrophage infiltration has also been demonstrated by antisense blockade of CSF-1 expression in colon cancer xenografts (Aharinejad et al 2002). CSF-1 stimulates the proliferation and differentiation of mononuclear cells and enhances MMP2 production in receptor-positive macrophages. CSF-1 blockade reduced tumour growth and vascularity, inhibited macrophage proliferation and recruitment and decreased tumour levels of MMP2 and angiogenic factors (Aharinejad et al 2002). As well as macrophage-derived gene products, a number of tumour-derived cytokines are associated with macrophage infiltration including IL6, MIP-1α, MCP-1, VEGF and GM-CSF. Many studies investigating the effects of over-expression of such cytokines have shown xenograft rejection due to activation of host-dependent antitumour responses. Studies have also shown increased macrophage infiltration associated with either marked tumour necrosis (Hoshino et al 1995) or larger numbers of well vascularised metastases (Di Carlo et al 1997) in response to overexpression of MCP-1 and IL6, respectively.

Hypoxia-regulated attractants and migration inhibitors

Macrophages can be attracted to an area by following a concentration gradient of a range of soluble chemotactic factors, many of which are expressed in various human tumours. Early studies noted that soluble factors were released in tumours of the ovary, skin and colon, some of which were stimulatory towards monocyte chemotaxis, and some inhibitory (Dammacco et al 1979, Bottazzi et al 1985, Hermanowicz et al 1987).

Hypoxia is known to regulate the expression of many genes in cancer cells through stabilization of the HIF transcription factor. Similar hypoxic up-regulation of HIF target genes has been demonstrated in primary human macrophages (Griffiths et al 2000). This partially explains the ability of macrophages to adapt to hypoxia by increased expression of glycolytic enzymes and their pro-angiogenic phenotype due to release of HIF-regulated cytokines such as VEGF and PDGF (Lewis et al 1999). A number of HIF-independent hypoxia-regulated genes also contribute to the phenotype of hypoxic macrophages with roles in angiogenesis (e.g. FGF2, TNFα), modulation of vascular tone (e.g. iNOS) and regulation of macrophage activation (e.g. MIP-1α), differentiation (e.g. GM-CSF), extravasation/migration (e.g. CD18 integrin, uPAR) and chemotaxis (e.g. MIP-1α, IL8; reviewed by Lewis et al 1999).

The initial recruitment of macrophages (and other immune cells) into areas of hypoxic tumour tissue is mediated by hypoxia-regulated genes expressed in other stromal cells and in cancer cells themselves. Adhesion of circulating monocytes to the vascular wall and subsequent transendothelial migration is enhanced under hypoxia by increased expression of endothelial cell adhesion molecules such as PECAM-1 (Kalra et al 1996). Macrophages are then induced to migrate into tumour tissue along gradients of hypoxia-regulated chemotactic agents such as VEGF, IL8, MIP-1α, endothelin 2 and MCP-1 (induced by PDGF and TNFα which are themselves hypoxia-regulated). Positive feedback mechanisms then serve to amplify the chemotactic signal. VEGF, for example, is expressed under hypoxia in both TAMs and tumour cells and exerts a chemotactic action on other macrophages, aiding their migration to avascular tumour sites (Lewis et al 2000). This suggests that hypoxia may be involved in the homing of macrophages into avascular regions of tumours whereupon they are activated to exhibit their own hypoxia response that may involve the direct stimulation of angiogenesis (Leek et al 1999, Leek et al 2002).

A complementary and/or alternative mechanism whereby macrophages are induced to accumulate in regions of tumour hypoxia involves the hypoxia-regulated inhibition of macrophage chemotaxis. This is partly mediated by increased expression of cytokines which inhibit macrophage migration, such as MIF (Takahashi et al 2001). More importantly however, the migration induced by a large number of chemoattractants seems to be directly inhibited by hypoxia itself. MCP-1- and RANTES-induced migration requires MAPK phosphorylation. Hypoxia rapidly increases the expression of MAPK phosphatase 1 (MKP-1) at the mRNA and protein level resulting in dephosphorylation of MAPK and inhibition of MCP-1-induced chemotaxis (Negus et al 1998). A similar mechanism has been described for the hypoxic inhibition of endothelin 2-mediated chemotaxis (Grimshaw et al 2002). These mechanisms combine to produce a rapid, specific response to hypoxia resulting in the selective accumulation of TAMs in avascular tumour sites.

Macrophages and anticancer therapy

From the above review it is clear that macrophages are contributors to tumour growth, invasion and metastasis, often through angiogenic pathways. It is therefore appropriate to consider that macrophages are a potential target for therapy. Amongst the mechanisms used by macrophages the following have already been investigated in pre-clinical or clinical studies with antagonists or modifiers for anticancer effects: VEGF, FGF2, EGF receptors, TNFα antagonism, thymidine phosphorylase inhibitors, HGF, kringle domains, IL8 antibodies, IGF binding protein, urokinase inhibitors and metalloprotease inhibitors. Thus directly or indirectly the macrophage will be a key target in anticancer therapy and often the major site of production of some of the above products depending on the individual tumour type.

However, these drugs are not specifically aimed at the macrophage and macrophages do provide other opportunities for specifically targeted therapy. While they may be involved in antigen presentation and of value in immunotherapy, the other possibility is to use them to target tumours directly because of their ability to migrate into tumour tissue.

Clinical protocols have been used involving the transfer of patient-derived macrophages that have been matured from peripheral monocytes and activated with interferon *in vivo* before administration. These have reported only very few side effects such as slight fever and chills. As yet this has not demonstrated antitumour activity and the effects could be related to several factors including dosing and inadequate localization.

This protocol has now been developed into a relatively straightforward clinical kit, which makes it available for randomized prospective studies. Exploiting this further with recent advances in gene transfer is an attractive possibility to consider, using the macrophages to deliver gene therapy at tumour sites throughout the body. A step in this direction has been achieved by using hypoxia response elements that are activated by the transcription factors HIF-1 and HIF-2 to regulate an enzyme that activates the pro-drug cyclophosphamide. Human cytochrome P450 2B6 linked to a hypoxia response element was transduced to macrophages and those macrophages were able to migrate into the centre of tumour spheroids and switch on the gene expression. Only those spheroids exposed to the transduced macrophages died when exposed to cyclophosphamide, demonstrating the principle of the approach.

Profiling of tumours is likely to become much more important in the future because anti-macrophage therapy or strategies relying on macrophage infiltration into tumours will not work unless the tumours are using these particular pathways. It is clear that detailed profiling of tumours, perhaps by gene arrays, will be necessary to select helpful therapies for patients. It is important to carry out clinical pathology studies relating the careful assessment of these components to the overall behaviour of the tumour and outcome.

References

Aharinejad S, Abraham D, Paulus P et al 2002 Colony-stimulating factor-1 antisense treatment suppresses growth of human tumour xenografts in mice. Cancer Res 62:5317–5324
Beasley NJ, Prevo R, Banerji S et al 2002 Intratumoral lymphangiogenesis and lymph node metastasis in head and neck cancer. Cancer Res 62:1315–1320

Bingle L, Brown NJ, Lewis CE 2002 The role of tumour-associated macrophages in tumour progression: implications for new anticancer therapies. J Pathol 196:254–265

Bottazzi B, Ghezzi P, Taraboletti G et al 1985 Tumor-derived chemotactic factor(s) from human ovarian carcinoma: evidence for a role in the regulation of macrophage content of neoplastic tissues. Int J Cancer 36:167–173

Brown NS, Jones A, Fujiyama C, Harris AL, Bicknell R 2000 Thymidine phosphorylase induces carcinoma cell oxidative stress and promotes secretion of angiogenic factors. Cancer Res 60:6298–6302

Dammacco F, Miglietta A, Lospalluti M, Meneghini C, Bonomo L 1979 Macrophages in skin cancer: quantitative and functional studies. Tumori 65:309–316

Di Carlo E, Modesti A, Castrilli G et al 1997 Interleukin 6 gene-transfected mouse mammary adenocarcinoma: tumour cell growth and metastatic potential. J Pathol 182:76–85

Elgert KD, Alleva DG, Mullins DW 1998 Tumor-induced immune dysfunction: the macrophage connection. J Leukoc Biol 64:275–290

Griffiths L, Binley K, Iqball S et al 2000 The macrophage—a novel system to deliver gene therapy to pathological hypoxia. Gene Ther 7:255–262

Grimshaw MJ, Balkwill FR 2001 Inhibition of monocyte and macrophage chemotaxis by hypoxia and inflammation—a potential mechanism. Eur J Immunol 31:480–489

Grimshaw MJ, Wilson JL, Balkwill FR 2002 Endothelin-2 is a macrophage chemoattractant: implications for macrophage distribution in tumours. Eur J Immunol 32:2393–2400

Hermanowicz A, Gibson PR, Jewell DP 1987 Tumour related inhibition of macrophage chemotaxis in patients with colon cancer. Gut 28:416–422

Hoshino Y, Hatake K, Kasahara T et al 1995 Monocyte chemoattractant protein-1 stimulates tumour necrosis and recruitment of macrophages into tumors in tumor-bearing nude mice: increased granulocyte and macrophage progenitors in murine bone marrow. Exp Hematol 23:1035–1039

Huang S, Van Arsdall M, Tedjarati S et al 2002 Contributions of stromal metalloproteinase-9 to angiogenesis and growth of human ovarian carcinoma in mice. J Natl Cancer Inst 94:1134–1142

Kaklamanis L, Leek R, Koukourakis M, Gatter KC, Harris AL 1995 Loss of transporter in antigen processing 1 transport protein and major histocompatibility complex class I molecules in metastatic versus primary breast cancer. Cancer Res 55:5191–5194

Kalra VK, Shen Y, Sultana C, Rattan V 1996 Hypoxia induced PECAM-1 phosphorylation and transendothelial migration of monocytes. Am J Physiol 271:H2025–H2034

Leek RD, Lewis CE, Whitehouse R, Greenall M, Clarke J, Harris AL 1996 Association of macrophage infiltration with angiogenesis and prognosis in invasive breast carcinoma. Cancer Res 56:4625–4629

Leek RD, Lewis CE, Harris AL 1997 The role of macrophages in tumour angiogenesis. In: Bicknell R, Lewis CE (eds) Tumour angiogenesis. Oxford University Press, Oxford

Leek RD, Landers R, Fox SB, Ng F, Harris AL, Lewis CE 1998 Association of tumour necrosis factor alpha and its receptors with thymidine phosphorylase expression in invasive breast carcinoma. Br J Cancer 77:2246–2251

Leek RD, Landers RJ, Harris AL, Lewis CE 1999 Necrosis correlates with high vascular density and focal macrophage infiltration in invasive carcinoma of the breast. Br J Cancer 79:991–995

Leek RD, Hunt NC, Landers RJ, Lewis CE, Royds JA, Harris AL 2000 Macrophage infiltration is associated with VEGF and EGFR expression in breast cancer. J Pathol 190:430–436

Leek RD, Talks KL, Pezzella F et al 2002 Relation of hypoxia-inducible factor-2 alpha (HIF-2 alpha) expression in tumor-infiltrative macrophages to tumor angiogenesis and the oxidative thymidine phosphorylase pathway in Human breast cancer. Cancer Res 62:1326–1329

Lewis JS, Lee JA, Underwood JCE, Harris AL, Lewis CE 1999 Macrophage responses to hypoxia, relevance to disease mechanisms. J Leukoc Biol 66:889–900

Lewis JS, Landers RJ, Underwood JC, Harris AL, Lewis CE 2000 Expression of vascular endothelial growth factor by macrophages is up-regulated in poorly vascularised areas of breast carcinomas. J Pathol 192:150

Negus RP, Turner L, Burke F, Balkwill FR 1998 Hypoxia down-regulates MCP-1 expression: implications for macrophage distribution in tumors. J Leukoc Biol 63:758–765

Schoppmann SF, Birner P, Stockl J et al 2002 Tumor-associated macrophages express lymphatic endothelial growth factors and are related to peritumoral lymphangiogenesis. Am J Pathol 161:947–956

Steele RJ, Eremin O, Brown M, Hawkins RA 1984 A high macrophage content in human breast cancer is not associated with favourable prognostic factors. Br J Surg 71:456–458

Sunderkotter C, Steinbrink K, Goebeler M, Bhardwaj R, Sorg C 1994 Macrophages and angiogenesis. J Leukoc Biol 55:410–422

Takahashi M, Nishihira J, Shimpo M et al 2001 Macrophage migration inhibitory factor as a redox-sensitive cytokine in cardiac myocytes. Cardiovasc Res 52:438–445

Talks KL, Turley H, Gatter KC et al 2000 The expression and distribution of the hypoxia-inducible factors HIF-1alpha and HIF-2alpha in normal human tissues, cancers, and tumor-associated macrophages. Am J Pathol 157:411–421

Wilbanks GD, Ahn MC, Beck DA, Braun DP 1999 Tumor cytotoxicity of peritoneal macrophages and peripheral blood monocytes from patients with ovarian, endometrial, and cervical cancer. Int J Gynecol Cancer 9:427–432

Williams CSM, Leek RD, Robson AM et al 2003 Absence of lymphangiogenesis and intratumoural lymph vessels in human metastatic breast cancer. J Pathol 200:195–206

DISCUSSION

Smyth: I have a general, conceptual comment about the use of this type of information for therapy of cancer. The maintenance of normoxia must be one of the most fundamental processes in the body. We have so many mechanisms that compensate if oxygen tension goes down. Understanding the biology of this is fascinating, but with the myriad of alternative ways of responding to the change in oxygen tension, do you think it will ever be possible to influence this in a xenobiotic way?

Harris: There is enough background to show that if we knockout HIF-1 in cell lines, it markedly inhibits the ability of tumours to grow in several different models. If you transfect into a cell line a domain of CBP300, with which HIF-1 interacts to effect transcription, this blocks tumour growth.

Smyth: I don't dispute that, but what about in the living mammal?

Harris: If you look at normal tissue distribution, there is hardly any HIF-1 around normally. We would exclude patients with a history of transient ischaemic attacks or myocardial infarction, as you would for any angiogenic inhibitor. I suppose the worry would be the effects of chronic use, and whether this would block pathways that are critical for survival. We don't know, but obviously VEGF inhibitors have been put into clinical trials with some success in renal cancer without horrendous tissue toxicity. You would probably combine these things with other therapies and not use them long term.

Smyth: My comment was more about the alternative ways in which the body can respond to a therapeutic manipulation to alter oxygen tension. It seems to me that we are so well protected.

Harris: HIF-1 knockout is fatal in embryogenesis, so there is no redundancy in mice.

Pollard: From your results, which were quite fascinating, the tumours take several strategies, but the end points are always the same, involving TP or VEGF, for example. There are several cell types which can be triggered in that way. Do you have any feeling for why a tumour would choose one route rather than another? Is there any sort of oncogenic activational pattern in those particular tumours that express TP in the epithelial cells, as opposed to those that don't?

Harris: You are asking an important question. If Fran Balkwill were talking about ovarian cancer, she would show different subcellular localization of the same factors, and Bob Strieter would show different localization in lung. Similar factors are important, and are expressed, but may have a different distribution. There was a nice study on transgenic models of mouse breast cancer presented at AACR last year. This involves three different models, and they assayed the endothelial cells from the tumours. They all responded slightly differently to angiogenic factors and expressed different signalling pathways on their surfaces. These different sorts of tumours have different sorts of blood vessels. This is reflected in the tissue specificity of oncogene action. Why does VHL give kidney cancer and not breast cancer? This could be an explanation. We suspect that each tissue has a specific pattern of being built. There is a common pattern, but the fine detail will vary.

Pollard: Going back to this idea of cancer recapitulating development, if we look at the way these particular tissues are built developmentally, will we see the same sort of heterogeneity of patterns?

Harris: I think this may well turn out to be the case. I like your model of breast cancer. It fits in with the idea of tissue specificity of oncogene action and gene function. If we look at the normal developmental pathway, this might well be what is diverted pathologically. Mina Bissell's work on stromelysins involves a similar concept.

Thun: The work you describe concerns surgically resected human breast cancers. The obvious strength of this is that the work is relevant to human cancer, and you can look at prognostic markers. I assume that these are all adjusted for stage.

Harris: They are all new primary tumours. The patients have had lymph node dissection and have all had either mastectomy or radiotherapy. If they are over 50, or oestrogen-receptor positive, they will all have had tamoxifen treatment.

Thun: You mentioned the issue of timing. You get cross-sectional slices, and not a dynamic longitudinal picture. How can you couple this work with animal models to get a dynamic picture?

Harris: This is the sort of thing that Jeff Pollard has been talking about. He came from the observation in humans of CSF-1 being important, and then carried out animal experiments that fitted extremely well with this. We found this EGFR association with macrophages some two years ago. The logical next step was to explore whether EGF or other cytokines from macrophages could help evolve different tumour pathways. You could imagine mixing experiments with 1% of the tumour cells having overexpressed EGFR, putting them into an animal model and then finding whether the EGFR overexpressors predominate a month or two later, and whether this could be modified by macrophages with different cytokine profiles. This might suggest that it is worthwhile giving inhibitors early on in the course of disease to try to prevent that tumour evolution happening.

Pollard: We are in the process of publishing a study where we have fairly carefully looked at some of the prognostic factors in human breast cancer in the polyoma middle T model. It is quite striking that there is down-regulation of oestrogen receptor and progesterone receptor, and up-regulation of cyclin D1 and HER-2/neu compared with normal tissue in this polyoma model. This is very stage dependent, so we can follow it. We should be able to use the timespan of the development of these sorts of cancers in the animal model, which are fairly reproducible, to be able to gain an understanding of the critical timing of those events. In addition, we may uncover the molecular mechanism behind them. No one really knows why the oestrogen receptor goes off or cyclin D1 goes up in breast cancers, for example.

Ristimäki: Could you stratify your material so that you could exclusively look in small tumours? In this subgroup the macrophages might predict the outcome.

Harris: No, we only have 130 cases. We are just trying to take a consecutive series with long-term follow up. The multivariate analysis we do is for size, grade, lymph node positivity, oestrogen-receptor-positivity and so on, but the numbers would be too small if we tried to look at tumours of less than 1 cm. We would need about 100 of those alone to do this. So far we haven't seen a clear association of size with any of the markers we have looked at.

D'Incalci: Can you explain the proline hydroxylase story in more detail?

Harris: There are three proline hydroxylases that modify HIF. One is hypoxia inducible, the other two aren't. The enzyme modifies two different proline residues on the HIF itself, producing hydroxyl proline. These sites can be recognized by the VHL protein, which is part of a ubiquitin ligase family. This ubiquitinates the HIF, which is then destroyed in the proteasome. The more proline hydroxylase is present, the better you are at destroying HIF. If you inhibit proline hydroxylase, which is an iron-requiring enzyme, the HIF goes up. If desferrioxamine is given to cells, it chelates iron, inhibits proline hydroxylase and HIF goes up dramatically. We think the hypoxia regulatable proline hydroxylase might be going down in the macrophages, and therefore the HIF-2 is going up.

Pepper: You mentioned that hypoxia increased telomerase activity in endothelial cells. Could you speculate on what you think this might mean?

Harris: These weren't immortal cells, they were Human Umbilical Vein Endothelial Cells (HUVECs). This tells us that if endothelial cells are quiescent cells, lying there doing nothing for 10 years, and then they suddenly need to expand their number, telomerase would be helpful for this process. What is the most likely stimulus for endothelial cells from trauma, infection or myocardial infarction? It will likely be hypoxia. This makes sense.

Pepper: I should point out that it has been shown that telomerase has anti-apoptotic activity which is independent of its ability to interact with telomeres. This may be as important for tumour cells as it is for endothelial cells.

Harris: It might be very important for endothelial cell migration. It would be interesting to turn off telomerase in endothelial cells and see whether it increases the apoptotic rate. Of course, there are inhibitors of these available.

Balkwill: I have a general methodology question. How could we best get macrophages out of tumours without activating them or re-educating them? It seems to me that by the time we have collagenase-digested a tumour, we have probably exposed every cell in it to endotoxin, and what we get out is not what was in there originally. Is there a magic trick, or will there be?

Mantovani: I haven't got it!

Harris: People do it, but I don't think they get very many out.

Balkwill: This is my point. If we look in tumours we see a lot of macrophages, but we don't get them out.

Harris: You could entice them out by using cytokines.

Pollard: The idea of needle collection is that we are very close. The caveat is that the needle collection itself may change the phenotype of the cell. This is one approach that is fairly proximal, though. Another approach is that we have been using GFP-labelled macrophages.

Balkwill: You can't do that in humans!

Pollard: It would probably be difficult to do the needle collection in humans. Laser capture is another option.

Harris: We want living cells.

Balkwill: I guess you could do laser capture and then amplify your RNA.

Harris: Claire Lewis (Sheffield Medical School) is getting out real numbers of macrophages from breast cancers and is trying to look at them.

Smyth: Give the macrophages iron filings and use a magnet!

Harris: There is a new imaging agent for macrophages that works on that principle using magnetic resonance imaging.

Balkwill: The techniques that Claire uses still involve a lengthy digestion of the tissue.

Pollard: It's the old problem of tissue culture. As soon as you culture a cell, it has changed.

Gordon: What about organ culture systems?

Pollard: They are better.

Balkwill: Marc Feldmann has been doing some interesting work in rheumatoid arthritis. He has indium-labelled the leukocytes of rheumatoid patients and shown that Remicade stopped the trafficking of cells.

Pollard: We have now done this in our breast cancer model. As a proof of principle we took GFP-labelled macrophages from the CSF-1 receptor knockout and put these back into the tumour. This was with cell lines, but there is no reason why you can't do this with primary cells. We left them for 30 minutes and then imaged them *in vivo* with a multi-photon microscope. The wild-type macrophages put into the tumours spread and move, and the CSF-1 receptor knockout macrophages sit there as round balls. You could always argue that the wild-type macrophages don't go into the tumour normally, but we hope that this is going to be one way that we can interfere with signalling. We would like to use a caging system *in vivo* so that we can actually uncage molecules in a cell *in vivo*.

Gordon: We have to try to get as much information as possible from direct *in situ* analysis using immunochemistry. Do you think the tools are OK for this?

Harris: For what we want to do, yes. One limiting factor is that there is a long list of chemokines, and for many of these we lack good antibodies that will work on paraffin-embedded sections or even frozen sections.

Strieter: We would be more than happy to provide you with those antibodies.

Gordon: A further difficulty will be getting hold of human tissue without falling foul of all the regulations.

Pollard: It is the same in the USA. It's a disaster.

The role of inflammation in tumour growth and tumour suppression

Thomas Blankenstein

Max-Delbrück-Centrum for Molecular Medicine, Robert-Rössle Strasse 10, 13092 Berlin and Institute of Immunology, Free University Berlin, Hindenburgdamm 30, 12200 Berlin, Germany

Abstract. The relationship between inflammation and tumour growth is poorly understood. The quality, quantity and time point of the inflammatory response may decide whether inflammation supports or inhibits tumour growth. Three examples are given that illustrate the different role of inflammation for tumour growth. It will be shown that tumour infiltrating macrophages can contribute to tumour rejection, can be essential for tumour growth or can occur as innocent bystander cells in tumours. Then it will be shown that the timely arrival of T cells at the tumour site is critical for tumour rejection and that non-bone marrow-derived tumour stromal cells are important targets during tumour rejection. Finally, a protective inflammatory response against the chemical carcinogen methylcholanthrene (MCA) will be discussed. This response is related to a tissue repair response induced by the tissue damaging effects of the carcinogen in the course of which MCA is encapsulated and no longer able to induce tumours.

2004 Cancer and inflammation. Wiley, Chichester (Novartis Foundation Symposium 256) p 205–214

The important relationship between cancer and inflammation has been recognized for some time. For a general overview some excellent reviews are recommended (Schreiber & Rowley 1999, Balkwill & Mantovani 2001, Coussens & Werb 2001). It has become apparent that immune cells can inhibit tumour growth and, paradoxically, stimulate tumour growth as originally proposed by Prehn (1994). Solid tumours usually contain normal non-malignant host cells, collectively termed the tumour stroma. The tumour stroma is built by inflammatory cells and often has a complex composition of macrophages/monocytes, granulocytes, T cells, fibroblasts (including their products, extracellular matrix protein), endothelial cells and other cells. The stromal content and composition can differ dramatically from one tumour to another and in relation to tumour progression. It is poorly understood at what time during tumour progression and which stromal cells are actively recruited by the tumour. The presence of inflammatory cells in the tumour *per se* gives no information of the role of these cells for tumour growth. They could be pro-tumorigenic (i.e. support tumour growth), or anti-tumorigenic (i.e. inhibit tumour growth), but inflammatory cells could also occur as innocent bystander cells without influencing tumour growth. The interaction between different tumour stromal cells may be important for tumour growth (Ibe et al 2001). I will give an example of how macrophage infiltration into the tumour can be modulated and how macrophage infiltration can determine whether a tumour grows or is rejected.

Rejection of a transplanted tumour usually requires T cells, either CD8[+] or CD4[+] or both. Their activation, in a major histocompatibility complex (MHC) class I or class II restricted fashion, requires antigen cross-presentation by antigen-presenting cells (APCs) in draining lymph nodes in most systems. Following activation T cells infiltrate the tumour and mediate their effector function. If the T cells arrive at the tumour site too late, tumour infiltration is impaired or effector functions are ineffective in eradicating the tumour (Blankenstein & Qin 2003). As discussed below, one reason is that interferon (IFN)γ, a key effector molecule produced by T cells, needs to be present at the tumour site during vascularization of the tumour.

IFNγ also plays a role in inhibition of tumour formation induced by 3-methylcholanthrene (MCA), since tumour incidence is increased in IFNγ receptor (IFNγR) gene-deficient (knockout; KO) compared to control IFNγR-competent mice (Qin et al 2002). This observation could have been interpreted as a phenomenon related to immunosurveillance as proposed by Burnet (1970). He suggested that one function of T cells is the recognition and elimination of tumour cells. However, analysis of the local events at the site of MCA suggests that the mechanism through which tumour development is suppressed in IFNγR-competent mice is different from T cell-mediated immunosurveillance and that under certain experimental conditions inflammatory responses can have anti-tumour effects in primary tumour models.

Changing the inflammatory response by expression of cytokines in tumour cells

A large number of experiments have been performed with tumour cells transfected to secrete a given cytokine (Blankenstein et al 1996). Usually, this does not alter the growth kinetics of the cells *in vitro*. Upon injection into mice the tumour cells secrete the cytokine locally so that cytokine-specific inflammatory cells in the tumour can be analysed. Furthermore, this assay allows the analysis of whether secretion of a particular cytokine influences tumour growth. Some examples illustrate the different results obtained with different cytokines with specific emphasis on the role of macrophages.

Expression of cytokines interleukin (IL)2, IL4, IL7, tumour necrosis factor (TNF), lymphotoxin (LT) and IFNγ in the same tumour cell line always led to rejection of the tumour cells upon injection into immunocompetent mice (Hock et al 1993, Qin & Blankenstein 1995). With the exception of IL7-secreting tumours tumour rejection was biphasic: long-term suppression occurred in the absence of T cells but complete tumour rejection required their presence. Rejection was cytokine-dependent, since neutralizing antibodies could restore tumour growth. Tumour rejection was preceded by an inflammatory response that was partially redundant and partially specific for the particular cytokine expressed by the tumour cells. For example, macrophages accumulated in larger numbers and more rapidly in all different cytokine-producing tumour cells compared to the parental tumour but eosinophils accumulated specifically in IL4- and IL7-secreting tumours (Hock et al 1991, 1993). At least in some cases the macrophages appeared to contribute to tumour rejection, since an antibody that inhibited macrophage migration at least partially restored tumour growth of TNF- or IL7-secreting tumour cells (Blankenstein et al 1991, Hock et al 1991). However, in general it is difficult to clearly demonstrate a cause–effect relationship between macrophage infiltration and tumour rejection. For instance, cytokine-producing tumour cells may die because of other reasons, perhaps because the cytokine inhibits tumour stroma formation, and macrophage infiltration is a secondary event as it occurs at sites of cell death or tissue damage. Several of the above cytokines, however, activate macrophages *in vitro* suggesting that in some situations macrophages can contribute to tumour rejection.

An interesting inverse relationship between tumour infiltrating macrophages (TIMs) and tumour growth has been observed with tumour cells transfected to secrete IL10, an anti-inflammatory cytokine (Richter et al 1993). Chinese hamster ovary (CHO) cells grew in T cell-deficient nude mice but their IL10-producing derivatives were rejected. Immunohistochemical analysis of the tumour tissue showed that the parental, progressively growing, tumour contained large numbers of TIMs, whereas in the IL10-producing tumour that was rejected TIMs were unable to infiltrate the tumour tissue and accumulated in the area surrounding the tumour. This suggested that the tumour needed to attract macrophages in order to grow and is compatible with the fact that tumours often secrete macrophage chemoattractants (Mantovani et al 1992). The type of help that the tumour received from TIMs is not known but an involvement of TIMs in neo-angiogenesis was proposed (Richter et al 1993).

Again a different result was obtained when colony-stimulating factor 1 (CSF-1, also known as M-CSF) was expressed in tumour cells by gene transfer (Dorsch et al 1993). The CSF-1-producing tumours contained large amounts of TIM but grew with the same kinetics compared to the parental tumour that was infiltrated by fewer numbers of macrophages and at a later time point.

These three examples illustrate that inflammatory cells in the tumour can be in different activation states and may have different, even opposite, effects on tumour growth. TIMs can contribute to tumour rejection, they can be necessary for tumour growth and they can occur as innocent bystander cells.

Tumour stromal cells as target during tumour rejection

If tumour cells are injected into mice, inflammatory cells infiltrate the tumour and form the stroma that the growing tumour needs. Recent experiments have shown that tumour stromal cells are the critical targets during T cell-mediated tumour rejection. IFNγ- and IFNγR-KO mice have a defect in tumour rejection. IFNγR-KO mice, in contrast to wild-type mice, could not reject a tumour challenge following immunization with irradiated tumour cells (Qin & Blankenstein 2000). Previous immunization was important, since naïve wild-type mice did not reject the tumour. The tumour cells did not need to express the IFNγR demonstrating that tumour rejection required host cells to respond to IFNγ. Adoptive T cell transfer and bone marrow chimera experiments revealed that T cells from immunized IFNγR-KO mice could reject tumours in wild-type mice and that IFNγR expression on non-bone marrow-derived cells was necessary for tumour rejection. Mice expressing IFNγR only on bone marrow-derived cells could not reject tumours. In immunized wild-type mice tumour-induced angiogenesis was effectively inhibited, since endothelial (CD31$^+$) cells were arrested at the border between tumour and adjacent tissue, so that the growing tumour was deprived of a blood supply, became necrotic and was finally rejected (Qin & Blankenstein 2000). Complete tumour rejection may additionally involve direct killing mechanisms, however, this needs further investigation. Together, the result showed that T cells, in this case CD4$^+$, had to rapidly infiltrate the tumour (ensured by the immunization) and produce IFNγ in order to inhibit rapid tumour burden by inhibition of tumour-induced angiogenesis. These data also demonstrate that T cells primarily attack tumour stromal cells and that tumour rejection does not entirely rely on direct killing of the tumour cells.

A protective inflammatory response during MCA-induced tumour development

Compared to the role of inflammation in tumour transplantation experiments, a different role of inflammation is seen during MCA-induced tumour development. MCA is a chemical carcinogen that induces — probably random — mutations. The carcinogenic action of MCA is primarily local, since tumours (often fibrosarcomas) usually appear at the injection site. Compatible with the assumption that the majority of mutations decrease cell survival rather than contribute to malignant transformation, dramatic tissue damage was detected at sites of MCA administration (Qin et al 2002). An inflammatory response typical for a tissue repair response with granulocytes, macrophages and large numbers of fibroblasts was immunohistochemically detected at sites of MCA injection. Importantly, encapsulated MCA was detected in long-term tumour-free mice. Therefore, we argued that a foreign body reaction in the course of a tissue-repair response protects from MCA-induced tumour development. IFNγR-KO mice were more susceptible to tumour development induced by MCA compared to normal control littermates (Qin et al 2002). Spontaneous tumour development in p53-KO mice was not increased in the absence of IFNγR expression suggesting that the protective role of IFNγ is specific for MCA but a role in other forms of tumour development cannot be excluded. Since local IFNγ production by tumour cells induced their encapsulation, we suggested that IFNγ contributed, by a poorly understood mechanism, to the tissue repair/encapsulation response against MCA. An unresolved question is where IFNγ comes from. The role of T cells in protection from MCA-induced tumour development is still controversial. However, the analysis of the local effects in the MCA-treated mouse suggests that the IFNγ-dependent response is primarily directed against MCA and not the tumour. This distinguishes the protective effect against MCA from immunosurveillance that is thought to act on tumour cells and to be mediated by T cells (Burnet 1970). Future studies have to show whether and which role T cells have during inhibition of MCA-induced tumours.

Conclusions

The relationship between inflammation and cancer is still poorly understood and will become one of the most important areas in tumour immunology. There is increasing evidence that inflammatory responses can stimulate but also inhibit tumour growth. It will be important to analyse in the future how pro- and anti-tumour inflammatory responses differ from each other. The role of inflammation needs to be analysed more physiologically, e.g. primary tumour, models, even though tumour transplantation models give useful information. Methods have to be found that convert a pro-tumour into an anti-tumour inflammatory response.

Acknowledgements

Work described here was supported by grants from the Deutsche Krebshilfe, Dr Mildred Scheel-Stiftung and the Deutsche Forschungsgemeinschaft.

References

Balkwill F, Mantovani A 2001 Inflammation and cancer: back to Virchow? Lancet 357:539–545
Blankenstein T, Qin Z 2003 The role of IFN-γ in tumor transplantation immunity and inhibition of chemical carcinogenesis. Curr Opin Immunol 15:148–154

Blankenstein T, Qin Z, Überla K et al 1991 Tumor suppression after tumor cell targeted tumor necrosis factor alpha gene transfer. J Exp Med 173:1047–1052

Blankenstein T, Cayeux S, Qin Z 1996 Genetic approaches to cancer immunotherapy. Rev Physiol Biochem Pharmacol 129:1–49

Burnet FM 1970 The concept of immunological surveillance. Progr Exp Tumor Res 13:1–27

Coussens LM, Werb Z 2001 Inflammatory cells and cancer: think different! J Exp Med 193:F23–F26

Dorsch M, Hock H, Kunzendorf U, Diamantstein T, Blankenstein T 1993 Macrophage colony-stimulating factor gene transfer into tumor cells induces macrophage infiltration but not tumor suppression. Eur J Immunol 23:186–190

Hock H, Dorsch M, Diamantstein T, Blankenstein T 1991 Interleukin 7 induces CD4+ T cell-dependent tumor rejection. J Exp Med 174:1291–1298

Hock H, Dorsch M, Kunzendorf U, Qin Z, Diamantstein T, Blankenstein T 1993 Mechanisms of rejection induced by tumor cell targeted gene transfer of interleukin 2, interleukin 4, interleukin 7, tumor necrosis factor or interferon-gamma. Proc Natl Acad Sci USA 90:2774–2778

Ibe S, Qin Z, Schüler T, Preiss S, Blankenstein T 2001 Tumor rejection by disturbing tumor stroma cell interactions. J Exp Med 194:1549–1559

Mantovani A, Botazzi B, Colotta F, Sozzani S, Ruco L 1992 The origin and function of tumor-associated macrophages. Immunol Today 13:265–270

Prehn RT 1994 Stimulatory effects of immune reactions upon the growths of untransplanted tumors. Cancer Res 54:908–914

Qin Z, Blankenstein T 1995 Tumor growth inhibition mediated by lymphotoxin: evidence of B lymphocyte involvement in the antitumor response. Cancer Res 55:4747–4751

Qin Z, Blankenstein T 2000 CD4+ T cell mediated tumor rejection involves inhibition of angiogenesis that is dependent on IFNγ receptor expression by nonhematopoietic cells. Immunity 12:677–686

Qin Z, Kim HJ, Hemme J, Blankenstein T 2002 Inhibition of methylcholanthrene-induced carcinogenesis by an interferon gamma receptor-dependent foreign body reaction. J Exp Med 195:1479–1490

Richter G, Krüger-Krasagakes S, Hein G et al 1993 Interleukin 10 gene transfected into Chinese hamster ovary cells prevents tumor growth and macrophage infiltration. Cancer Res 53:4134–4137

Schreiber H, Rowley DA 1999 Inflammation and cancer. In: Gallin JI, Snyderman R, Fearon DT et al (eds) Inflammation: basic principles and clinical correlates. 3rd edn, Lippincott Williams & Wilkins, Philadelphia, p 1117–1129

DISCUSSION

Oppenheim: One ancient definition of cancer is as a wound that fails to heal. This is clearly a bit simplistic. I really don't understand your pessimism about the IFNγ receptor. You say that it promotes encapsulation and inhibits the diffusion of MCA. You pointed out that this was a wonderful thing: when you deposited the material into muscle or subcutaneously it controlled the response and caused a foreign body reaction. So why not say that IFNγ is very protective in that situation?

Blankenstein: I have never said anything else. What I have said is that it is problematic to equate this with immune surveillance. Immune surveillance was postulated by Burnet in order to find a function for T cells. We have no evidence for an involvement of T cells in this process. We have to make a clear distinction, therefore, between the foreign body reaction and Burnet's idea of immune surveillance. This has some implications. It suggests that you cannot compare MCA-induced tumours with spontaneous tumours, and you cannot explain the mechanism controlling tumour development in one tumour model by experiments in another tumour model, such as tumour transplantation experiments.

Brennan: In one of your diagrams there was a suggestion that IFNγ was coming from the macrophages. If it is not coming from the T cells or NK cells, where is it coming from?

Blankenstein: In this case, when we injected cyclophosphamide, IFNγ came from the macrophages. This may be surprising. A few years ago it was shown that macrophages produce large amounts of IFNγ when they are cultured with IL12 and IL18. We repeated these experiments with tumour-infiltrating macrophages, and found that they indeed produce IFNγ.

Brennan: That is very surprising to me.

Gordon: Although this work has been published, it is still quite controversial.

Blankenstein: Then let me be more specific. The tumour-infiltrating macrophages were sorted twice for Mac1-positive cells by MACS. The majority were F4/80-positive and were strongly plastic adherent.

Gordon: Mac1 is present on NK cells too.

Blankenstein: But not F4/80.

Gordon: You could have a subpopulation of non-macrophage cells in a mixture, and they may be the ones that are making IFNγ. Have you done *in situ* hybridization?

Blankenstein: No.

Gordon: So you have a mixed population.

Blankenstein: The population of cells we isolated before we treated with cyclophosphamide produced IL10. Then, in the tumour-bearing mice, if we gave cyclophosphamide and waited for 24 h, they had switched to produce IFNγ. However, I agree with you, that MACS sorting does not give a 100% homogeneous population.

Mantovani: I am intrigued by the linking of IFNγ with fibrosis. There is evidence that IFNγ has antifibrotic activity. It has even been used in the clinic, and one paper claims that it works as an antifibrotic.

Strieter: It is contentious.

Mantovani: If you use bleomycin to induce lung fibrosis, IFNγ is protective. What is the mechanism for this? To me, it seems counterintuitive.

Blankenstein: If IFNγ-secreting tumours grow for a long period in T cell-deficient mice, they are finally encapsulated. I would not exclude that this is the end product of chronic blood deprivation, since IFNγ inhibits tumour-induced angiogenesis. I am not sure whether this is mediated directly by IFNγ. Therefore, it is possible that the fibrosis at sites of chronic IFNγ production is an indirect effect.

Cerundolo: In the first part of your talk you mentioned that CD4-positive cells have an effect on the tumours that are negative for class II. Since not all the CD4-positive cells are class II restricted, can you rule out the possibility that those CD4-positive cells are in fact CD1-restricted cells and that CD1restricted CD4$^+$ T cells may be capable of slowing down tumour progression?

Blankenstein: No, we have not done this, simply because I would be very surprised if systemic tumour immunity is conferred by CD1-restricted CD4 cells.

Forni: Here you have a model in which granuloma formation is a mechanism to prevent certain kinds of carcinogenesis. This new form of immunosurveillance differs from those already described. IFNγ is a key factor for granuloma formation. No IFNγ, no granuloma, in fact, as you showed with the tuberculosis reaction. We now have to find out what cells are involved in granuloma formation. Are T cells or NK cells involved? This is a different issue, but one that can be explored. NK cells or T cells probably are involved in granuloma formation, which means that what you have described is a completely new immunosurveillance mechanism that differs from those already described, even though the cells involved are the same.

Blankenstein: T cells are certainly involved in the response against tuberculosis and the response is antigen-specific. Granulomas contain fibrotic components. The question is whether the role of IFNγ during granuloma formation is to regulate immune responses or, perhaps supports, directly or indirectly, extracellular matrix deposition.

Thun: I have a question about the injected MCA model. It is a very different sort of exposure than one might get from polycyclic aromatic hydrocarbons in tobacco smoke, for example, in that it is all together in one place. With inhaled polycyclic aromatic hydrocarbons you get a widespread deposition throughout the bronchi and it goes on for a long time, against which the approach of encapsulating the carcinogen would seem to be less effective. You made a good case that encapsulation is one of the ways that the immune system works, but it might not be an effective way of protecting against inhaled chemical carcinogens.

Blankenstein: You may be right. This is difficult to test.

Balkwill: Following on from that, I didn't quite understand your comment about two-stage carcinogenesis. One of the classic ways to induce papillomas on the skin of mice is to use DMBA as the initiating carcinogen and then, instead of using TPA as a tumour promoter, induce a wound in the mice. The cells and cytokines produced during wound healing act as endogenous tumour promoters.

Blankenstein: I wanted to say that the protective role of IFNγ is specific for MCA. This is suggested by preliminary experiments employing two-stage skin carcinogenesis.

Balkwill: That wouldn't surprise me at all.

Blankenstein: A tumour-promoting role of inflammation in two-stage carcinogenesis is well established. Therefore, it may be interesting that inflammation is protective during MCA-induced carcinogenesis.

Brennan: Other than doing these experiments in the IFNγ receptor knockouts, have you treated the wild-type mice with IFNγ to see whether there is a time-dependent involvement? In a model of inflammation, collagen-induced arthritis, expression of IFNγ early on is detrimental to the disease process, but expressed later on it is protective. So by blocking IFNγ we get the expected result of protection early but exacerbation later on. Interestingly, when we backcrossed IFNγ knockouts onto DBA mice, there was no difference in the incidence, onset or acceleration of arthritis compared with littermate controls. Obviously, knockouts are useful, but we can't see this time-dependent involvement of these molecules.

Blankenstein: I agree with you. That is a problem. To answer your question, IFNγ-neutralization *in vivo* is problematic, because you need enormous amounts of antibodies, since you have to treat the mice for a very long time. Additionally, antibodies against the anti-IFNγ antibody could impair the results. In knockout mice it is difficult to exclude developmental defects which contribute to the phenotype.

Oppenheim: In a sense, you bring up an issue that is crucial. We have spent the last few days saying that under certain circumstances inflammation can be associated with and promote cancer, and even perhaps be causal. So why under some circumstances is inflammation not doing this? The question becomes what kind of inflammation, and how much of it, does it take for the development of cancer?

Richmond: Perhaps it has to do with the genetic background: a little bit of inflammation might be pro-tumorigenic in one person and not in another, because of a variety of differences.

Balkwill: One of the problems is defining inflammation in the first place.

Thun: One issue raised earlier was the difference between spontaneous tumours and acquired, exogenously-produced tumours. In humans it is hard to make a clear distinction. Only for certain childhood cancers, which are thought to be two-stage, is there much expectation that risk is predominantly inherited rather than acquired? Even in childhood cancers, an individual may inherit one hit but need to acquire the other. The fact that one doesn't know how the other hit is acquired doesn't mean that the risk of developing the cancer is inherited rather than sporadic (acquired). If we look at adult tumours, most cases involve a mix of inherited and acquired factors. Lung cancer risk comes predominantly from exogenous carcinogens in tobacco smoke. Breast and prostate cancer are highly influenced by exposure to endogenous hormones. In humans it is extremely hard to break down what is spontaneous versus what is acquired. Furthermore, a lot of the acquiring is from endogenous processes, like hormone cycling. I don't think that this is a distinction that we can be too dogmatic about at this point.

Gordon: I think Richard Doll would take a similar view: that much of it is environmental and we just don't know the details.

Oppenheim: There is an interesting phenomenon that we ignore as scientists. There is a common observation that before the Second World War, stomach and upper gastrointestinal (GI) tract cancer was quite common in the west, as it still is in Japan and the Far East. Since then, there has been a tremendous drop, which correlates with the advent of refrigeration. You could propose that all the pickles and preserved foods that are still eaten in the Far East may be causing inflammation in the stomach and upper GI tract. *Helicobacter* may be favoured by this. By changing our lifestyle in the west we may have moved the cancer from the upper GI to the lower GI tract. Now we have to figure how to move it out all together!

Cyclooxygenase 2: from inflammation to carcinogenesis

Ari Ristimäki

Department of Pathology, Helsinki University Central Hospital and Molecular and Cancer Biology Research Program, Biomedicum Helsinki, University of Helsinki, Finland

Abstract. Cyclooxygenase (COX) is the rate-limiting enzyme in the conversion of arachidonic acid to prostanoids. Two COX isoforms have been cloned, of which COX-1 is constitutively expressed, while the expression of COX-2 is low or nondetectable in most tissues, but can be readily induced in response to cell activation by cytokines, growth factors and tumour promoters. Thus, COX-1 is considered a housekeeping gene and thought to be responsible for the synthesis of prostanoids involved in cytoprotection of the stomach and for the production of the pro-aggregatory prostanoid thromboxane by the platelets. In contrast, COX-2 is an inducible, immediate-early gene, and its role has been related to inflammation, reproduction and carcinogenesis. Expression of COX-2 is elevated in a variety of human malignancies and in their precursor lesions. Furthermore, genetic deletion or pharmacological inhibition of COX-2 suppresses tumour growth in several animal models of carcinogenesis. In humans, elevated COX-2 expression is associated with poor prognosis in adenocarcinomas of the digestive tract and the breast, and a selective inhibitor of COX-2 reduced polyp burden in patients who suffer from familial adenomatous polyposis. Thus, COX-2 seems to be a relevant target in chemoprevention.

2004 Cancer and inflammation. Wiley, Chichester (Novartis Foundation Symposium 256) p 215–226

COX-2 and inflammation

Cyclooxygenase (COX) is the rate-limiting enzyme in the conversion of arachidonic acid to prostanoids (Vane et al 1998). Two COX isoforms have been cloned (COX-1 and COX-2) that share over 60% identity at amino acid level and have similar enzymatic activities (DuBois et al 1998, Taketo 1998a). The most striking difference between the two COX enzymes is in the regulation of their expression. While COX-1 is constitutively expressed, the expression of COX-2 is low or not detectable in most tissues, but can be readily induced in response to cell activation by hormones, proinflammatory cytokines, growth factors and tumour promoters. Up-regulation of COX-2 expression by proinflammatory cytokines

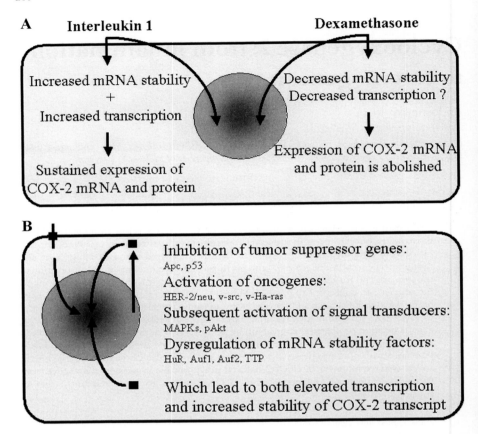

A **Interleukin 1**   **Dexamethasone**

Increased mRNA stability
+
Increased transcription

Decreased mRNA stability
Decreased transcription ?

Sustained expression of
COX-2 mRNA and protein

Expression of COX-2 mRNA
and protein is abolished

B

Inhibition of tumor suppressor genes:
Apc, p53

Activation of oncogenes:
HER-2/neu, v-src, v-Ha-ras

Subsequent activation of signal transducers:
MAPKs, pAkt

Dysregulation of mRNA stability factors:
HuR, Auf1, Auf2, TTP

Which lead to both elevated transcription
and increased stability of COX-2 transcript

FIG. 1. (A) Regulation of COX-2 expression in non-neoplastic cells (fibroblasts, macrophages and endothelial cells) is primarily regulated by para- or autocrine factors such as cytokines, growth factors and hormones. (B) Regulation of COX-2 expression in malignant cells is primarily regulated by intracrine phenomena such as dysregulation of oncogenes (activating mutations or gene amplification) or tumour suppressor genes (inactivating mutations or gene deletions).

(Ristimäki et al 1994) and its down-regulation by anti-inflammatory glucocorticoids (Ristimäki et al 1996) is, at least in part, due to modulation of the transcript stability (Fig. 1A). Since COX-2 is expressed in rheumatoid arthritis (Crofford et al 1994) but not in normal gastrointestinal tract (Kargman et al 1996), it was hypothesized that COX-2-selective inhibitors could relieve inflammatory symptoms without inducing the side-effects of conventional non-steroidal anti-inflammatory drugs (NSAIDs). Thus, COX-1 is considered a housekeeping gene and thought to be responsible for the synthesis of prostanoids involved in cytoprotection of the stomach, vasodilatation in the kidney and for production of the pro-aggregatory prostanoid thromboxane by the platelets. In contrast, COX-2 is an inducible, immediate-early gene, and its role has been related to inflammation, reproduction and carcinogenesis (DuBois et al 1998, Taketo 1998b).

COX-2-selective inhibitors have proved to be effective drugs in the treatment of symptoms of rheumatoid arthritic and osteoarthritic patients (Bombardier et al 2000, Silverstein et al 2000). In addition, they seem to cause less adverse gastrointestinal side effects than conventional NSAIDs. Thus, patients that have major risk factors for NSAID-induced side effects, such as advanced age and history of peptic ulcer disease, should benefit from the use of COX-2 selective inhibitors over the conventional NSAIDs. However, the above described COX-1/COX-2 hypothesis may prove to be too simplistic. It needs to be realized that, in addition to COX-1, COX-2 is also expressed in the brain, possibly in the vascular endothelium, and certainly in the kidney. Especially cardiovascular issues need to be studied more carefully (elevation of blood pressure, frequency of myocardial events and combination treatment with low-dose aspirin). There exists one study (VIGOR), in which a significant increase of myocardial infarction was observed among the patients using a COX-2-selective drug (rofecoxib) against the comparator (naproxen) (Bombardier et al 2000). The cause of adverse cardiovascular events remains unresolved, and no definitive conclusion can be drawn based on present data. It is important to note that no conclusions can be drawn by comparing this specific study to any other study (for example due to the differences in patient selection and combination with other treatments such as aspirin). Possible explanations for the results of the VIGOR study include (in alphabetical order): chance, naproxen is cardioprotective, and/or cardiovascular side-effects are associated with use of rofecoxib. We clearly need more careful clinical trials focused on these cardiovascular issues.

Next generation COX-2 inhibitors (etoricoxib and valdecoxib) have been introduced. It is not clear whether these drugs provide any advantages over the first generation drugs (celecoxib and rofecoxib). In fact, it is unlikely that these drugs would prove to be more effective than the older ones (and, unfortunately, it is very likely that no head to head comparisons will be performed). However, since pharmacokinetic (rather than selectivity) properties may play an important role in the side-effect profile, it is important to perform careful clinical trials bearing in mind the cardiovascular issues.

COX-2 in carcinogenesis

Epidemiological studies indicate that the use of NSAIDs associates with reduced risk of malignancies especially in the digestive tract (Thun et al 2002). Expression

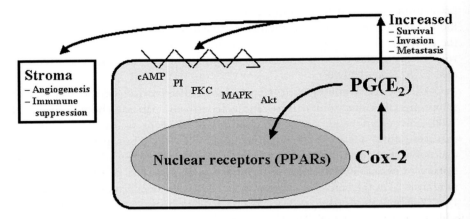

FIG. 2.   Overexpression of COX-2 has been linked to increased resistance to apoptosis of the cancer cells. COX-2 products also enhance production of angiogenic factors (VEGF), increase invasion and metastasis (possibly via matrix metalloproteinases) and act as immunosuppressive agents.

of COX-2 is elevated in a variety of human carcinomas, and inhibition of COX-2 suppresses tumour growth in several animal models of carcinogenesis (Dannenberg et al 2001, Gupta & DuBois 2001, Van Rees & Ristimäki 2001). The mechanism by which COX-2 is up-regulated in cancers is unknown, but one possibility is that cancer cells become intrinsically more active in expressing COX-2 than do the non-neoplastic cells. To this end, both inactivation of tumour suppressor genes, such as p53, and activation of oncogenes, such as HER-2/neu, have been implicated in induction of COX-2 expression (Fig. 1B). Functionally, COX-2-derived prostanoids have been shown to promote angiogenesis, induce invasion and increase metastasis (Fig. 2). Probably the most convincing data on the direct role of COX-2 in carcinogenesis are based on studies, which show that genetic deletion of COX-2 leads to a reduced number and size of polyps in a mouse model of familial adenomatous polyposis (FAP) (Oshima et al 1996) and that overexpression of COX-2 promote tumour formation in the mammary gland in a transgenic mouse model (Liu et al 2001). Thus, <u>COX-2 seems to be a relevant target in chemoprevention</u>. Indeed, several ongoing clinical trials are investigating the chemopreventive or therapeutic (adjuvant) effects of COX-2 selective inhibitors. The first one of these trials has been published, which indicated that celecoxib (a selective COX-2 inhibitor) reduced polyp burden in patients who suffer from FAP (Steinbach et al 2000).

We have previously characterized COX-2 expression in several human malignancies and in their precursor lesions (Ristimäki et al 1997, Wolff et al 1998,

Ristimäki et al 2001, Saukkonen et al 2001, Van Rees et al 2002). Our data suggest that COX-2 is primarily expressed in neoplastic epithelial cells of human carcinomas, and that both preinvasive and invasive cancer cells express COX-2. However, metaplastic, and thus non-neoplastic, precursor lesions of gastric and oesophageal adenocarcinoma express only low levels of COX-2. We have recently studied COX-2 expression in oesophageal, breast and ovarian carcinomas, and in each case elevated COX-2 expression correlated with poor prognosis:

- In oesophageal adenocarcinoma expression of COX-2 was elevated in 79% of the tumours, and patients with high COX-2 expression were more likely to develop distant metastases and local recurrences and their survival was significantly reduced. Furthermore, expression of COX-2 was recognized as an independent prognostic factor by multivariate analysis (Buskens et al 2002).
- In tissue array specimens of 1576 invasive breast cancers (FinProg material) elevated expression of COX-2 protein was observed in 37% of the tumours, and it was associated with unfavourable distant disease-free survival (Ristimäki et al 2002). Interestingly, COX-2 was a more significant prognostic factor in the tumours that had good prognosis characteristics (based on oestrogen receptor, p53 and HER-2/neu status).
- We have also analysed expression of COX-2 protein in human serous ovarian carcinoma material. Elevated expression of COX-2 was detected in 70% of the tumours, and it was associated with poor differentiation and with reduced survival. Interestingly, elevated expression of COX-2 correlated with both aberrant p53 and aberrant SMAD4 immunostaining. This suggests that dysregulated expression of tumour suppressor genes may be the cause for up-regulation of COX-2 in serous ovarian adenocarcinomas (Erkinheimo et al 2003).

In accord with these data elevated COX-2 expression was shown to associate with poor prognosis in colorectal carcinomas as well (Sheehan et al 1999). Thus, COX-2 may be an important tumour promoter in several different types of human adenocarcinomas. However, the prognostic significance of COX-2 expression may differ in a single organ site based on the (histological and/or genetic) subtype of the tumour. To this end, it is important to know in which types of malignancies COX-2 is expressed and how it affects the prognosis of the patient. These studies will help to target the use of COX-2 inhibitors as chemopreventive and therapeutic agents to those patients who will most likely benefit from this treatment.

Acknowledgements

The Helsinki University Central Hospital Research Funds, the Finnish Cancer Foundation and the Academy of Finland supported the original work.

References

Bombardier C, Laine L, Reicin A et al 2000 Comparison of upper gastrointestinal toxicity of rofecoxib and naproxen in patients with rheumatoid arthritis. VIGOR Study Group. N Engl J Med 343:1520–1528

Buskens CJ, Van Rees BP, Sivula A et al 2002 Prognostic significance of elevated cyclooxygenase 2 expression in patients with adenocarcinoma of the esophagus. Gastroenterology 122:1800–1807

Crofford LS, Wilder RL, Ristimäki AP et al 1994 Cyclooxygenase-1 and -2 expression in rheumatoid synovial tissues: Effects of interleukin-1β, phorbol ester, and corticosteroids. J Clin Invest 93:1095–1101

Dannenberg AJ, Altorki NK, Boyle JO et al 2001 Cyclo-oxygenase 2: a pharmacological target for the prevention of cancer. Lancet Oncol 2:544–551

DuBois RN, Abramson SB, Crofford L et al 1998 Cyclooxygenase in biology and disease. FASEB J 12:1063–1073

Gupta RA, Dubois RN 2001 Colorectal cancer prevention and treatment by inhibition of cyclooxygenase-2. Nat Rev Cancer 1:11–21

Erkinheimo TL, Lassus H, Finne P et al 2003 Elevated cyclooxygenase-2 expression associates with aberrant p53 and SMAD4 and poor outcome in serious ovarian carcinoma. Clin Cancer Res, in press

Kargman S, Charleson S, Cartwright M et al 1996 Characterization of prostaglandin G/H synthase 1 and 2 in rat, dog, monkey, and human gastrointestinal tracts. Gastroenterology 111:445–454

Liu CH, Chang SH, Narko K et al 2001 Overexpression of cyclooxygenase-2 is sufficient to induce tumorigenesis in transgenic mice. J Biol Chem 276:18563–18569

Oshima M, Dinchuk JE, Kargman SL et al 1996 Suppression of intestinal polyposis in Apc delta716 knockout mice by inhibition of cyclooxygenase 2 (COX-2). Cell 87:803–809

Ristimäki A, Garfinkel S, Wessendorf J, Maciag T, Hla T 1994 Induction of cyclooxygenase-2 by interleukin-1alpha: evidence for a post-transcriptional mechanism. J Biol Chem 269:11769–11775

Ristimäki A, Narko K, Hla T 1996 Down-regulation of cytokine-induced cyclooxygenase-2 transcript isoforms by dexamethasone: evidence for post-transcriptional regulation. Biochem J 318:325–331

Ristimäki A, Honkanen N, Jänkälä H, Sipponen P, Härkönen M 1997 Expression of cyclooxygenase-2 in human gastric carcinoma. Cancer Res 57:1276–1280

Ristimäki A, Nieminen O, Saukkonen K, Hotakainen K, Nordling S, Haglund C 2001 Expression of cyclooxygenase-2 in human transitional cell carcinoma of the urinary bladder. Am J Pathol 158:849–853

Ristimäki A, Sivula A, Lundin J et al 2002 Prognostic significance of elevated cyclooxygenase-2 expression in breast cancer. Cancer Res 62:632–635

Saukkonen K, Nieminen O, van Rees B et al 2001 Expression of cyclooxygenase-2 in dysplasia of the stomach and in intestinal type gastric adenocarcinoma. Clin Cancer Res 7:1923–1931

Sheehan KM, Sheahan K, O'Donoghue DP et al 1999 The relationship between cyclooxygenase-2 expression and colorectal cancer. J Am Med Assoc 282:1254–1257

Silverstein FE, Faich G, Goldstein JL et al 2000 Gastrointestinal toxicity with celecoxib vs nonsteroidal anti-inflammatory drugs for osteoarthritis and rheumatoid arthritis: the CLASS study: A randomized controlled trial. Celecoxib Long-term Arthritis Safety Study. J Am Med Assoc 284:1247–1255

Steinbach G, Lynch PM, Phillips RK et al 2000 The effect of celecoxib, a cyclooxygenase-2 inhibitor, in familial adenomatous polyposis. N Engl J Med 342:1946–1952

Taketo MM 1998a Cyclooxygenase-2 inhibitors in tumorigenesis (part I). J Natl Cancer Inst
90:1529–1536

Taketo MM 1998b Cyclooxygenase-2 inhibitors in tumorigenesis (Part II). J Natl Cancer Inst
90:1609–1620

Thun MJ, Henley SJ, Patrono C 2002 Nonsteroidal anti-inflammatory drugs as anticancer
agents: mechanistic, pharmacologic, and clinical issues. J Natl Cancer Inst 94:252–266

Van Rees BP, Ristimäki A 2001 Cyclooxygenase-2 in carcinogenesis of the gastrointestinal tract.
Scand J Gastroenterol 36:897–903

Van Rees BP, Saukkonen K, Ristimäki A et al 2002 Cyclooxygenase-2 expression during
carcinogenesis in the human stomach. J Pathol 196:171–179

Vane JR, Bakhle YS, Botting RM 1998 Cyclooxygenases 1 and 2. Annu Rev Pharmacol Toxicol
38:97–120

Wolff H, Saukkonen K, Anttila S, Karjalainen A, Vainio H, Ristimäki A 1998 Expression of
cyclooxygenase-2 in human lung carcinoma. Cancer Res 58:4997–5001

DISCUSSION

Thun: Earlier in this meeting, Siamon Gordon said he was somewhat sceptical about whether this field was ready to go. I was thinking about this. The exception may be in the area of clinical trials of the selective COX-2 inhibitors in all these different kinds of cancers. Ari Ristimäki, which of these show greatest promise?

Ristimäki: The field is clearly concentrating on colorectal carcinogenesis. As with all clinical trials in cancer biology, it is difficult to predict what will happen. The matrix metalloproteinase (MMP) field is a good example of this.

Thun: One of the things that makes me a little worried is that a lot of these clinical trials go into play without really understanding the stage of tumorigenesis that is likely to be affected, and without good intermediate markers. This is particularly true in the colon cancer trials. The trial of aspirin that John Baron's group completed found a paradoxical effect of a reduction in the occurrence of sporadic colorectal polyps in the low-dose group of aspirin (a baby aspirin a day), and not in the higher dose group (an adult aspirin a day). Neither of these doses would produce sustained inhibition of COX-2 in a nucleated cell, but particularly not the baby aspirin. These surprises keep appearing, which is a source of worry.

Ristimäki: There are published data that not only by knocking out COX-2, but also by knocking out COX-1 you can affect tumorigenesis. It might be that COX-1 is a player, at least in rodents, in the very early stages of tumorigenesis. Perhaps the baby aspirin has an effect on these early lesions also in humans. Along with the large clinical trials, it is our responsibility to concentrate on smaller trials and try to figure out the biomarkers that are modulated in human settings. We know a lot about tissue culture and animal models, but we need to know more about the human situation. This should also help direct trials to the populations that will benefit the most. I am very happy about our breast cancer trial (Ristimäki et al 2002). The way that these breast cancer trials were conducted signified a significant change of policy. The data show that we will most likely see an effect in oestrogen receptor (ER)-positive tumours when using COX-2 selective inhibitors. At first the industry didn't want to do anything more than just treat breast cancer as a single entity, but now they are also conducting trials targeted to patients with ER-positive tumours. I think academic research can have significance in redirecting trials and showing that treating molecularly characterised subpopulations within the same disease, rather than treating the disease as a whole, can lead to more significant results.

Wilkins: We have heard a lot in the last few days about the importance of macrophages in tumour development. Has anyone looked at COX-2 inhibition effects on macrophage populations in these tumours?

Ristimäki: Probably not so much in tumours. But that is probably the target in rheumatoid arthritis, for example. In respect of immunosuppression in the cancer situation, it may well be that it is both the tumor cells and the macrophage that people are targeting when they inhibit prostaglandin E$_2$-mediated immune suppression by using COX-2 selective drugs. It is a viable hypothesis.

Mantovani: What is downstream of COX-2? It is logical to assume that it is prostaglandins. But from there on, there are several prostaglandin receptors. On different macrophage populations, for example, it is not as simple as prostaglandin E$_2$ being suppressive. The opposite is true for brain macrophages: it will amplify tumour necrosis factor (TNF) production, because these cells have a different prostaglandin receptor repertoire. Is there unequivocal evidence about the key downstream receptor in the pro-tumour function of COX-2? Has anyone tested prostaglandin receptor knockout mice, for example?

Ristimäki: I didn't have time to go into the receptors in my talk. As you know, there are four different receptors for prostaglandin E$_2$, and some of these have multiple splice variants. It is a very complicated field. In the APC knockout model, EP$_2$ knockout gives the same kind of tumour suppressor effect as knocking out COX-2. So EP$_2$ seems to be a relevant receptor in the FAP model. I know that many groups are studying the role of PPAR gamma and delta in tumour formation. Since these nuclear receptors are also targets for both prostanoids and NSAIDs, it is not only the membrane receptors that we need to consider, but also the nuclear receptors.

Strieter: I wanted to add an additional comment in terms of a potential mechanism. In collaboration with Steve Dubinett at UCLA we have looked specifically at NSCLC lung cancer cell lines which overexpress COX-2. We have studied the interrelationship with their potential phenotype in terms of proangiogenesis. This is highly associated with the expression of ELR-positive CXC chemokines. If you attenuate COX-2 this attenuates the angiogenic phenotype through the ELR-positive CXC chemokines. The other downstream mechanism for this is an alteration of redox potential of the cell, leading to translocation of NF-κB and the activation of a variety of NF-κB-associated genes. This is another potential mechanism in which COX-2 may very well be important without subsequent autocrine effects through prostaglandin E$_2$. This is an effect through what is essentially a prostaglandin E$_2$-independent mechanism.

Oppenheim: You mentioned that there is a change in the ratio of IL10:IL12 activity. I didn't catch what the change was, but this may be the key to what is going on.

Strieter: This has also been demonstrated in work from Dubinett's group. They have shown that prostaglandin E$_2$ can shift the balance in favour of over-expression of IL10 as compared with IL12.

Oppenheim: That would relate to Alberto Mantovani's question. It would support the idea that a type II macrophage phenotype would be favoured by COX-2 expression by tumours.

Strieter: That is IL10 expression from a tumour cell, not a macrophage.

Oppenheim: It would influence the behaviour of neighbouring macrophages.

Caux: Besides macrophages, there is a clear effect of prostaglandin E$_2$ on dendritic cells. Prostaglandin E$_2$ blocks IL12 on dendritic cells and causes IL10 production by macrophages. Perhaps this is an additional part of the prostaglandin E$_2$ effect.

Thun: I was intrigued by the discussions yesterday about the infiltration of macrophages to various degrees in many breast tumours and other tumours. Perhaps characterizing COX-2 in relation to macrophage infiltration could be informative, as well as characterizing the type of macrophage.

D'Incalci: I am very interested by the observation that the abundance of COX-2 can be a prognostic factor for survival only in those cases that have a more differentiated tumour. Do you think there is no hope for undertaking a conventional clinical trial with advanced-stage tumours, and instead that it would be better to concentrate clinical investigation on early phases of development?

Ristimäki: The field of clinical research is concentrating on chemoprevention rather than treatment. Treating existing tumours with COX inhibitors alone is not a feasible strategy. At best, they can be used in an adjuvant setting in conjunction with something like kinase inhibitors of the EGFR family, radiotherapy or chemotherapy. There again, you may need to focus on certain subtypes of tumours in order to see the benefit. My speculation is that we would need to understand the make-up of the tumour well before we could expect to see a response.

D'Incalci: The determinants for efficacy aren't known. You say that we have to characterize the tumour, but do you know what to measure in order to predict efficacy?

Ristimäki: It would be nice to see COX-2 expression in the tumours. In breast tumours we see significant expression in less than half of the tumours. This might be a starting point. Others might argue that it doesn't matter because they think that COX-2 is in the stroma, and every solid tumour has a stroma. With our methods we don't see that much COX-2 in the stroma when we compare this with the neoplastic epithelial cells, not even in the blood vessels. This might be dependent on the sensitivity of our assay.

Gordon: In terms of prevention, what are the risks in putting healthy people on COX-2 inhibitors?

Ristimäki: Two large trials have been published. The CLASS trial (Silverstein et al 2000) used celecoxib and showed a favourable response with respect to GI side-effects against the comparitor (ibuprofen) at a 6 month time point, but not at a 12 month time point. Only the FDA and the company have the conclusive data, so it not easy to make definitive conclusions. In the other trial, that I described in my paper (Bombardier et al 2000) there was an increase in the myocardial infarction rate in patients using rofecoxib when compared to naproxen. In that trial aspirin was not allowed, although 4% of this population of 8000 patients actually were in a high risk cardiovascular group and would have needed aspirin. It may be that this COX-2-selective inhibitor that may block circulating prostacyclin and elevate blood pressure by about 5 mm of mercury may actually throw the patients who are on the edge of myocardial infarction in an unfavourable direction. This is a concern, which needs to be clarified. A big question concerns the efficacy in the treatment of neoplasias. COX-2 inhibitors are known to be effective in respect to arthritis management, but are they as effective as non-selective drugs in prevention of human cancer deaths?

Thun: In chemoprevention, particularly in cancer, the events are low enough in average risk groups that the drugs have to be incredibly safe in order to have a favourable risk–benefit balance. Chemoprevention has pretty much focused on very high risk groups so that the trials can give informative results in a reasonable time frame for a reasonable cost, and also because the risk–benefit ratio is most likely to be favourable in high risk groups. A separate issue about the selective COX-2 inhibitors is that they haven't been used for long enough in large enough populations so that we fully understand their side effects in vulnerable subgroups of the population.

Oppenheim: I'd like to describe some data showing the interrelationship between COX-2 products, VEGF and CXCR4. Our initial observation was that stimulation of CXCR4 expression on endothelial cells was up-regulated very actively, in the range of 5–7-fold, by VEGF and FGF, which are the classical angiogenic factors. Conversely, SDF stimulation of CXCR4 increases the production of VEGF and FGF. If I had known what I learnt today from Ari Ristimäki, that prostaglandin acts as an angiogenic factor through VEGF, it would have made things a bit simpler. But we did learn that prostaglandin was involved. The question that we raised was whether the generation of prostaglandin by VEGF and FGF was influencing CXCR4. Indeed, in terms of migration effects as an indication of expression of CXCR4 and the ability of endothelial cells to migrate in response to SDF, there was considerable up-regulation by VEGF, FGF and prostaglandin of functional CXCR4 in comparison with controls. There was low level constitutive expression, but this is up-regulated by all these particular stimuli. Then we asked whether the COX inhibitors block the induction of VEGF and FGF, and expression of CXCR4. Even the basal expression is reduced to some extent by inhibitors such as aspirin and NS398. VEGF induction is markedly reduced by COX inhibitors, as is the FGF induction of CXCR4 expression. By using fluorescence to detect new blood vessels we have seen that prostaglandin induces increased vascular development. Prostaglandin increases expression of CXCR4 very nicely on these new capillaries in matrigel implanted in mice, and it is reduced by anti-CXCR4. Anti-CXCR4 reduces the capacity of prostaglandin to induce tubule formation in implanted matrigel plugs. There is therefore an interrelationship in which prostaglandin, induced by VEGF or FGF2, up-regulates CXCR4 expression. In our hands, at least, evaluation of prostaglandin effects on other chemokine receptors did not have this effect. This seemed to be unique for CXCR4, which is a developmental chemokine involved in angiogenesis in embryonic and fetal development. Thus, the angiogenic effects of prostaglandin are mediated to a considerable extent by CXCR4. In contrast, VEGF and FGF angiogenesis is only partially dependent on CXCR4, so there are other pathways through which they are angiogenic.

Mantovani: I have a general question for both the prostaglandin and CXCR4 people. Is there anything special in the responsiveness to prostaglandins or CXCR4 agonists in the gastrointestinal (GI) vasculature? The reason I raise this point is that the knockouts have a specific effect in the GI vasculature. Of course, we have heard everything about COX-2 and the GI system. Did you test GI endothelial cells? Does anyone have GI endothelial cells?

Pepper: Not as far as I know.

Gordon: It could be a good place to look.

Oppenheim: The cells we used were microvascular cells from tissue specimens or from circumcision specimens, or cell lines. It is a mystery why in the SDF/CXCR4 knockout there are selective defects in the gastrointestinal mesenteric area. We don't know whether there is anything special about GI endothelial cells.

Strieter: Yesterday we were looking at the issue of why you can have CXCR4 within a tumour but not necessarily have SDF1- and CXCR4-mediated angiogenesis. It is simply competition. If there is a significant expression of CXCR4, what you see on the tumour cells is that it out-competes the endothelium. This is an important consideration. But this is not necessarily true for other ligands that promote angiogenesis in the context of that tumour system. This is one of the reasons why there may well be a disparity between primary tumour angiogenesis related to other factors and SDF1/CXCR4-mediated angiogenesis.

Oppenheim: In line with this we have attempted to inhibit CXCR4/SDF1 interaction in mice to see whether this has an antitumour effect. The expression of CXCR4 is so widespread on tissues and tumours that the anti-CXCR4 was not sufficiently potent: we estimated it would have cost us US$80 000 to buy enough antibody to do this experiment properly.

Strieter: I can give you anti-SDF1 for a lot cheaper than this!

Oppenheim: We also tried T22, another inhibitor, which was not sufficient to inhibit. The experiments do suggest that CXCR4 may be a target and better inhibitors, which we are trying to obtain, may be effective. We certainly would like to test anti-SDF1.

Thun: One thing that is distinctive about the gut is that the loss of the *APC* gene is an early event in the malignant progression of at least colon and stomach cancer. I don't know how that epithelial event relates to what you are talking about in the endothelium, but it is a characteristic.

References

Bombardier C, Laine L, Reicin A et al 2000 Comparison of upper gastrointestinal toxicity of rofecoxib and naproxen in patients with rheumatoid arthritis. VIGOR Study Group. N Engl J Med 343:1520–1528

Ristimäki A, Sivula A, Lundin J et al 2002 Prognostic significance of elevated cyclooxygenase-2 expression in breast cancer. Cancer Res 62:632–635

Silverstein FE, Faich G, Goldstein JL et al 2000 Gastrointestinal toxicity with celecoxib vs nonsteroidal anti-inflammatory drugs for osteoarthritis and rheumatoid arthritis: the CLASS study: a randomized controlled trial. Celecoxib Long-term Arthritis Safety Study. J Am Med Assoc 284:1247–1255

The inflammatory cytokine network of epithelial cancer: therapeutic implications

Peter Szlosarek and Fran Balkwill[1]

Cancer Research UK, Translational Oncology Laboratory, Barts and The London, Queen Mary's School of Medicine and Dentistry, John Vane Science Centre, London EC1M 6BQ, UK

Abstract. A network of inflammatory cytokines and chemokines is found in epithelial cancer. There is evidence that these mediators contribute not only to tumour cell growth and survival, but also to communication between the cancer cells and stromal elements. Tumour cell production of the inflammatory cytokine tumour necrosis factor α (TNFα) is one critical factor in this autocrine and paracrine network. TNFα may also initiate and sustain production of other cytokines and chemokines. Chemokines are key determinants of the leukocyte infiltrate in solid ovarian tumours and influence the extent and phenotype of the infiltrate in ascitic disease. Chemokines may also be involved in tumour cell spread. Thus some of the processes involved in chronic inflammation are also active in human epithelial cancers. Agents that antagonize TNFα or chemokines are currently being assessed in preclinical animal cancer models and in phase I clinical trials.

2004 Cancer and inflammation. Wiley, Chichester (Novartis Foundation Symposium 256) p 227–240

The epithelial tumour microenvironment is a complex tissue comprising variable numbers of tumour cells, fibroblasts, endothelial cells and infiltrating leucocytes. Cytokines and chemokines are key molecules controlling autocrine or paracrine communications within and between these individual cell types.

Under some circumstances, these endogenous mediators may orchestrate host responses against the tumour, but there is increasing evidence that the cytokine/chemokine network contributes to tumour growth, progression and host immunosuppression. Hence, cytokines and chemokines influence growth and survival of the epithelial tumour cells; orchestrate development and remodelling of the extracellular matrix; regulate the leukocyte infiltrate; stimulate

---

[1]This paper was presented at the symposium by Fran Balkwill to whom all correspondence should be addressed.

neovascularization; manipulate the immune response; and may contribute to the multi-step process of metastasis.

This chapter will outline some of the actions of endogenous cytokines in epithelial tumours with particular emphasis on tumour necrosis factor α (TNFα) and chemokines. Inhibition of proinflammatory influences in a cancer may provide a useful addition to other biological therapies of cancer, allow optimal action of a specific immune therapy or be a useful addition to chemotherapy regimes. Moreover, drugs that are active in inflammatory disease may 'translate' into cancer therapies.

TNFα as a tumour promoter

Induction of TNFα by pathogenic stimuli induces cytokines, chemokines, endothelial adhesions and matrix metalloproteases that recruit and activate cells at sites of infection and tissue damage. In a therapeutic setting, TNFα is classically considered to have anti-cancer activity via stimulation of T cell mediated immunity and haemorrhagic necrosis, but administration is associated with severe toxicity. It is also clear that low amounts of TNFα produced chronically by cancers can have tumour promoting activity (Balkwill 2002).

Tumour promoting activity of TNFα in animal tumour models

There is evidence from a range of animal cancer models that TNFα can have tumour promoting activity. TNF null (TNFα$^{-/-}$) mice were resistant to skin carcinogenesis induced by DMBA initiation and TPA tumour promotion (Moore et al 1999). This resistance was related to a temporal delay in the activation of protein kinase Cα (PKCα) and AP-1, key TPA-responsive signalling molecules, leading to decreased expression of genes involved in tumour development (GM-CSF, MMP-3 and MMP-9), in TNFα$^{-/-}$ mice compared to wild-type animals (Arnott et al 2002). Mice deficient in the p55 TNF receptor, TNFRI$^{-/-}$ mice, and to a lesser extent p75 receptor null TNFRII$^{-/-}$ mice, were also resistant to skin carcinogenesis (our unpublished observations).

TNFRI$^{-/-}$ mice displayed reduced oval cell (hepatic stem cell) proliferation during the preneoplastic phase of liver carcinogenesis, and developed fewer tumours than wild-type mice (Knight et al 2000). Endogenous TNFα was also found to be critical in promoting liver metastasis following intrasplenic administration of colonic adenocarcinoma cells (Kitakata et al 2002). An increase in TNFα in the tumour also promotes cancer development and spread. For instance, experimental lung metastases were enhanced upon pre-treatment of the animals with TNFα (Orosz et al 1995) and overexpression of TNFα conferred invasive properties on xenograft tumours.

Expression of TNFα in the human tumour microenvironment

Tumour and/or stromal expression of TNFα has been reported in the following malignancies: breast, ovarian, colorectal, prostate, bladder, oesophageal, renal cell cancer, melanoma, lymphomas and leukaemias (Balkwill 2002). In ovarian cancer, for example, TNFα mRNA and protein were expressed mainly within tumour epithelial islands whereas in colorectal cancer predominant expression was found within the tumour associated macrophages. TNFRI was found on tumour and stromal cells, and TNFRII localized to the leukocyte infiltrate in ovarian cancer suggesting potential for both autocrine and paracrine actions of endogenous TNFα. In serous ovarian carcinoma, prostate carcinoma, non-Hodgkin's lymphoma, acute leukaemia, chronic myelogenous leukaemia and chronic lymphocytic leukaemia there was an association between expression of the cytokine, poor survival and/or resistance to therapy. Recent studies in gastro-oesophageal cancer revealed a progression in TNFα and TNFRI epithelial expression along the Barrett's metaplasia–dysplasia–carcinoma sequence. E-cadherin was negatively regulated by TNFα, and in common with several other epithelial cancers, was down-regulated in Barrett's adenocarcinoma, with corresponding elevations in cytoplasmic/nuclear pools of β catenin (Jankowski et al 2000, Tselepis et al 2002).

Table 1 summarizes the various mechanisms by which TNF-α may promote cancer growth, invasion and metastasis.

TNFα and paraneoplastic phenomena

Increasing tumour burden is characterized by progressive weight loss, with muscle and lipid wasting or cachexia. This syndrome is driven by several tumour and/or

TABLE 1 Tumour-promoting effects of TNFα

- Production of NO (DNA/enzyme damage, cGMP-mediated tumour promotion)
- Autocrine growth and survival factor for malignant cells
- Activation of E6/E7 mRNA in HPV-infected cells
- Tissue remodelling via induction of matrix metalloproteinases
- Control of leukocyte infiltration in tumours via modulation of chemokines and their receptors
- Down-regulation of E cadherin, increased nuclear pool of β catenin
- Enhance tumour cell motility and invasion
- Induction of angiogenic factors
- Loss of androgen responsiveness
- Resistance to cytotoxic drugs

host-derived factors including the cytokines TNFα, IL1, IL6 and IFNγ. Although animal data are convincing for the involvement of TNFα in cachexia, clinical studies evaluating the cytokine in the human syndrome have been less consistent (Tisdale 2001). However, the data for specific tumours, for example pancreatic carcinoma, are more robust with recent studies confirming both increased TNFα mRNA and protein in blood analysed from patients with cachexia. Other paraneoplastic features linked to TNF-α include the fever, hypercalcaemia, anaemia and fatigue typical of many tumour presentations with advanced disease. TNFα has been characterized as an endogenous pyrogen, osteoclast activating factor facilitating skeletal metastases, and as a suppressor of erythropoietin production.

TNFα and treatment-related complications

Several pro-inflammatory cytokines, such as TNFα and IL6, mediate cytokine release syndromes complicating, in particular, antibody-based cancer therapy. Studies of rituximab in patients with haematological malignancy confirmed TNFα release as a major biological event, correlating with fever, rigors and chills. A recent study also revealed a major role for TNFα in cisplatin-induced nephrotoxicity in a murine model. Although several proinflammatory cytokines and chemokines, including TGFβ, RANTES, MIP2 and MCP-1, were up-regulated during cisplatin-induced renal injury, the protection afforded by TNFα antagonists or deletion of the TNFα gene, demonstrated an absolute requirement for TNFα (Ramesh & Reeves 2002).

TNFα antagonists

Specific TNFα antagonists, such as infliximab and etanercept, are now accepted as successful treatments for a variety of inflammatory and autoimmune conditions. The behaviour of the inflammatory pannus of rheumatoid arthritis (RA) may be likened to the invasion and metastasis of cancer, with up-regulation of angiogenic factors, MMPs, cytokines and chemokines. In addition, studies in RA patients have shown that infliximab treatment reduced production of inflammatory cytokines, angiogenic factors and MMPs, inhibited leukocyte trafficking to inflamed joints and improved bone marrow function (Maini & Taylor 2000). All these actions could be useful in a biological therapy for cancer.

Reports of activity of TNFα antagonists in haematological cancers are beginning to emerge, with improvements in bone marrow function in myelodysplastic syndrome (MDS), myelofibrosis (MF) and acute myelogenous leukaemia, and a significant role in acute and chronic graft versus host disease (GVHD). A number of studies have now been initiated in various solid tumours using TNF-α antagonists either alone, or in combination with chemotherapy, and for the palliative treatment of cancer cachexia.

Non-specific TNFα inhibitors

Thalidomide, dexamethasone, aspirin and non-steroidal anti-inflammatory drugs (NSAIDs) can modulate both TNFα dependent and independent pathways. Dexamethasone, one of the first non-specific TNFα inhibitors clinically available, has widespread application in oncology. It inhibits both TNFα transcription and translation. Thalidomide also represses TNFα at several levels, promoting decay of the mRNA transcript and decreasing NF-κB binding activity. Several studies have confirmed that thalidomide has activity in a range of tumours, including multiple myeloma, renal cell and prostate carcinoma, melanoma, glioma and Kaposi's sarcoma (Eisen 2002). Patients have also reported improvements in symptoms of anorexia, weight loss, nausea, insomnia and profuse sweating in the palliative care setting.

There is good epidemiological evidence for a protective role of aspirin and NSAIDs against colorectal cancer, and emerging data suggesting benefit in a variety of other epithelial tumours (see Ristimäki 2003, this volume). Both groups of anti-inflammatory drugs inhibit cyclooxygenase 2 (COX-2), which is induced by TNFα and other inflammatory cytokines via a NF-κB pathway.

Chemokines and cancer

The network of chemokines and their receptors that is found in epithelial cancer may contribute to various aspects of tumour development and spread. Chemokines are major determinants of the leucocyte infiltrate in many cancers and may contribute to an immunosuppressive microenvironment. They also have positive and negative influences on angiogenesis, may act as growth and survival factors for both tumour and stromal cells, and stimulate tumour cell migration (Balkwill 2003, Balkwill & Mantovani 2001, Vicari & Caux 2002, Wilson & Balkwill 2002).

Chemokines and the leukocyte infiltrate

The presence of leukocytes in solid tumours is related to local production of chemokines by both tumour and stromal cells (Balkwill & Mantovani 2001, Bottazzi et al 1983). CC chemokines are major determinants of macrophage and lymphocyte infiltration, for instance, in human carcinomas of the breast and cervix, sarcomas and gliomas. In breast cancer, there is a positive correlation between macrophages, lymph node metastasis and clinical aggressiveness, and the tumour cells have been reported to produce the chemokines CCL2 (MCP-1) and CCL5 (RANTES) (Luboshits et al 1999). Within the tumour microenvironment of epithelial ovarian cancer, the predominant infiltrating cells are macrophages and CD8$^+$ T lymphocytes. CCL2 localized to epithelial areas of the tumour and correlated with the extent of this lymphocyte and macrophage infiltration (Negus et al 1997).

Chemokines contribute to immune suppression in tumours

Infiltrating leukocytes may not only contribute to tumour progression by producing MMPs and growth, angiogenic and immunosuppressive factors, but the profile of the cells attracted by chemokines to the tumour may contribute to an immunosuppressive environment (Balkwill & Mantovani 2001). There is often a prevalence of Th2 cells in tumours and this polarization may be a general strategy to subvert immune responses against tumours. Hodgkin's disease, for instance, is characterized by constitutive activation of NF-κB and overproduction of Th2 cytokines and chemokines into the tumour, causing an influx of Th2 cells and eosinophils. These reactive cells are thought not only to contribute to proliferation and survival of the malignant cells, but to suppression of cell-mediated immunity. Chronic exposure to high chemokine concentrations in the tumour microenvironment may encourage activated Type II macrophages that release immunosuppressive IL10 and TGFβ. These macrophages may also release CCL2, which could contribute to a Th2 polarized immunity and stimulate a type II inflammatory response (Sica et al 2000).

A balance of angiogenic and angiostatic chemokines exists in tumours

Chemokines may also regulate angiogenesis in the epithelial tumour microenvironment. CXC chemokines containing the three amino acid (ELR) motif glutamine-leucine-arginine promote angiogenesis (Belperio et al 2000). They are directly chemotactic for endothelial cells and can stimulate angiogenesis in neovascularization experiments *in vivo*. Elevated levels of CXCL5 were found, for instance, in primary non-small-cell lung cancer, NSCLC, and correlated with the vascularity of the tumours. In contrast, CXC chemokines such as CXCL9 and CXCL10, lack the ELR motif are often anti-angiogenic. Levels of CXCL10 in human lung cancers were inversely related to tumour progression.

Chemokines and tumour cell growth

Deregulated chemokines may contribute directly to transformation of tumour cells by acting as growth and survival factors, generally in an autocrine manner. This action of chemokines has been extensively characterized in malignant melanoma.

CXCL1 and CXCL8 are constitutively produced by melanoma cells, but not by untransformed melanocytes. Melanoma cells also show elevated levels of the CXCR2 receptor for these chemokines, and autocrine chemokine stimulation enhances survival, proliferation and tumour cell migration (Dhawan & Richmond 2002). In addition, CXCL8, CXCL1, as well as the CC chemokine CCL20 (MIP-3α), all stimulate growth of pancreatic tumour cell lines. In epithelial ovarian cancer, tumour cells express CXCR4 and its ligand CXCL12 enhances tumour cell proliferation in conditions of sub-optimal growth. This autocrine chemokine stimulation may also have paracrine implications because CXCL12 stimulated the production of TNFα by ovarian tumour cells (Balkwill 2002).

Malignant cells may respond to chemokine gradients

Mechanisms involved in leukocyte homing may also be used by tumour cells. Restricted and specific expression of chemokine receptors by tumour cells, especially CXCR4 and CCR7 may be one important step in the development of site specific metastasis. For example, tumour cells from breast, prostate, pancreatic, gastric and ovarian carcinomas, neuroblastoma, glioblastoma, melanoma and some leukaemias, express a limited and highly specific range of chemokine receptors (Balkwill 2003, Muller et al 2001, Scotton et al 2001). Generally, only one or two of the known chemokine receptors are expressed on each tumour type. In breast, prostate and ovarian cancer, neuroblastoma, melanoma and some forms of leukaemia, the respective ligand is strongly expressed at sites of tumour spread. Expression of CCR7 and CCR10 on melanoma cells was linked to expression of ligands for these receptors at the two major sites of metastasis, skin and lymph nodes (Muller et al 2001). Functional CCR7 and CXCR4 chemokine receptors are found on human breast cancer cells. The ligand for CCR7, CCL21, (SLC) is highly expressed in lymph nodes and in a tissue screen, strong expression of the ligand for CXCR4, CXCL12 (SDF1), was only seen in the target organs for breast cancer metastases. Moreover, breast cancer cells migrated towards tissue extracts from these target organs, and chemotaxis could be partially abrogated by neutralizing antibodies to CXCR4 (Muller et al 2001).

Of 14 CC and CXC chemokine receptors investigated, only CXCR4 was expressed on ovarian cancer cells (Scotton et al 2001). Its ligand CXCL12 (SDF1) was found in ovarian cancer ascites and in tumour biopsies, but neither receptor nor ligand was expressed by normal ovarian epithelium. CXCL12 stimulated tumour cell proliferation *in vitro* and migration/invasion towards a CXCL12 gradient. Stimulation of cells with CXCL12 resulted in sustained activation of Akt/PKB, biphasic phosphorylation of p44/42 MAPK and induction of mRNA and protein for the pro-inflammatory cytokine TNFα, a cytokine involved in tumour/stromal communication in ovarian cancer (Scotton et al 2002).

Animal cancer studies provide some experimental proof that cancer cells may co-opt normal mechanisms of leukocyte homing to lymph nodes. When B16 melanoma cells were transduced with a retroviral vector containing cDNA for the chemokine receptor CCR7, metastasis to lymph nodes was increased (Wiley et al 2001).

Manipulation of the chemokine network in tumours

In spite of extensive pre-clinical studies of chemokine and chemokine receptor antagonists in inflammatory and infectious disease models, there are little published data on such strategies in models of malignant disease.

Direct evidence for the involvement of chemokines in this stromal tumour promotion comes from studies of melanoma cell lines growing in nude mice (Nesbit et al 2001). The effects of xenograft tumour-derived CCL2 depended on the level of secretion. Low-level production of CCL2 was associated with modest macrophage infiltration, tumour formation and neovascularization of the tumour mass. High CCL2 secretion led to a massive macrophage infiltration and destruction of tumour cells. Attempts to decrease macrophage infiltration in tumours by neutralizing chemokines/receptors or chemokine antagonists, seem warranted.

Another approach is to overexpress certain chemokines in the tumour microenvironment (Vicari & Caux 2002). In the past ten years many experimental animal studies have shown that overexpression of chemokines can greatly increase tumour-associated leukocyte numbers and phenotypes, with concomitant stimulation of destructive immune or inflammatory responses. This was first shown by overexpressing CCL2 in experimental tumours and stimulating a destructive macrophage influx (Bottazzi et al 1992). Overexpression of CCL20 (MIP3α) suppressed tumour growth, by attracting dendritic cells to activate tumour-specific cytotoxic T lymphocytes (Fushimi et al 2000). CCL19 (MIP3β) overexpression mediated rejection of murine breast tumours in an NK cell and CD4+ T cell-dependent manner and CCL21 (6Ckine) reduced growth of a colon carcinoma cell line in mice using a similar mechanism (reviewed in Vicari & Caux 2002). To date, these delivery approaches have not been translated into clinical trial.

Chemokines in anti-angiogenic strategies

In experimental models using human non-small-cell lung carcinoma (NSCLC) cell lines grown in SCID mice, neutralization of CXCL5 and CXCL8 reduced tumour growth, vascularity and metastasis, but had no direct effect on *in vitro* cell growth. The ELR$^+$ chemokine CXCL8 was also important in promoting angiogenesis and tumorigenesis in human ovarian cancer xenografts implanted into the peritoneum of nude mice. In this study, mouse survival was inversely associated with CXCL8 expression within the tumour. Overexpression of the ELR$^-$ chemokines CXCL9 and CXCL10 inhibited the growth of Burkitt's lymphoma tumours established in nude mice. Likewise, administration of intra-tumoural CXCL10 inhibited NSCLC tumorigenesis and metastasis in SCID mice. CCL21 (6Ckine) also inhibited growth and angiogenesis of human lung cancer cell lines in SCID mice (for review see Balkwill 2003).

Chemokine antagonists and metastases

Chemokine and chemokine receptor antagonists may also have direct actions on tumour cells. In this respect there are three published studies of antibodies that neutralize chemokines or chemokine receptors. As described above, B16 melanoma cells transduced with a retroviral vector containing cDNA for the chemokine receptor CCR7, showed increased lymph node metastasis. Lymphatic spread of these CCR7 transfected cells was blocked by neutralizing antibodies to the CCR7 ligand, CCL21 (Wiley et al 2001). A majority of non-Hodgkin's lymphoma cells express CXCR4 and treatment of cells *in vitro* with neutralizing antibodies to CXCR4 inhibited cell migration, pseudopodia formation and decreased proliferation. In an animal model (NOD/SCID) of human high-grade NHL, CXCR4 neutralization delayed tumour growth, reduced the weight of any tumours that subsequently developed and significantly increased mouse survival. Neutralizing antibodies to CXCR4 also suppressed lymph node metastasis in a xenograft model of a CXCR4-expressing breast cancer (Bertolini et al 2002).

Conclusions

Chronic endogenous low level production of TNFα in tumours may contribute to the development of tissue architecture necessary for tumour growth and metastasis, induce other cytokines, angiogenic factors and MMPs, contribute to DNA damage and enhance growth/survival of tumour cells. Thus, TNFα provides a molecular bridge between inflammation and cancer. Based on the success of TNFα antagonists in inflammatory disease, further investigation of these therapeutics is warranted in combating cancer, complications of malignancy and the adverse events of conventional treatment.

The chemokine network in human tumours is complex and its role is only partially understood. There is a great deal of information on potential roles of individual chemokines, but more work is needed to define the overall chemokine and chemokine receptor profile of individual tumour types, and the cells within them. The chemokines that are produced by human tumours are part of an even more complex network of inflammatory, immunomodulating and growth promoting cytokines, and little is known about interactions between all these mediators. A common theme is emerging of the involvement of constitutively activated NF-κB in inflammatory cytokine and chemokine dysregulation in cancers.

Preclinical studies of chemokine antagonists in animal cancer models are warranted, as well as investigations of tumour development and spread in animals deficient in individual chemokines or receptors. Such experiments could provide a rationale for novel approaches to cancer treatment, where manipulation of the chemokine balance could be a useful adjunct to existing biological therapies of cancer.

References

Arnott CH, Scott KA, Moore RJ et al 2002 Tumour necrosis factor-alpha mediates tumour promotion via a PKCalpha- AP-1-dependent pathway. Oncogene 21:4728–4738

Balkwill F 2002 Tumor necrosis factor or tumor promoting factor? Cytokine Growth Factor Rev 13:135–141

Balkwill F 2003 Chemokine biology in cancer. Semin Immunol 15:49–55

Balkwill F, Mantovani A 2001 Inflammation and cancer: back to Virchow? Lancet 357:539–545

Belperio JA, Keane MP, Arenberg DA et al 2000 CXC chemokines in angiogenesis. J Leukoc Biol 68:1–8

Bertolini F, Dell'Agnola C, Mancuso P et al 2002 CXCR4 neutralization, a novel therapeutic approach for non-Hodgkin's lymphoma. Cancer Res 62:3106–3112

Bottazzi B, Polentarutti N, Acero R et al 1983 Regulation of the macrophage content of neoplasms by chemoattractants. Science 220:210–212

Bottazzi B, Walter S, Govoni D, Colotta F, Mantovani A 1992 Monocyte chemotactic cytokine gene transfer modulates macrophage infiltration, growth, and susceptibility to IL-2 therapy of a murine melanoma. J Immunol 148:1280–1285

Dhawan P, Richmond A 2002 Role of CXCL1 in tumorigenesis of melanoma. J Leukoc Biol 72:9–18

Eisen T 2002 Thalidomide in solid malignancies. J Clin Oncol 20:2607–2609

Fushimi T, Kojima A, Moore MAS, Crystal RG 2000 Macrophage inflammatory protein 3a transgene attracts dendritic cells to established murine tumors and suppresses tumor growth. J Clin Invest 105:1383–1393

Jankowski JA, Harrison RF, Perry I, Balkwill F, Tselepis C 2000 Barrett's metaplasia. Lancet 356:2079–2085

Kitakata H, Nemoto-Sasaki Y, Takahashi Y, Kondo T, Mai M, Mukaida N 2002 Essential roles of tumor necrosis factor receptor p55 in liver metastasis of intrasplenic administration of colon 26 cells. Cancer Res 62:6682–6687

Knight B, Yeoh GCT, Husk KL et al 2000 Impaired preneoplastic changes and liver tumor formation in tumor necrosis factor receptor type 1 knockout mice. J Exp Med 192:1809–1818

Luboshits G, Shina S, Kaplan O et al 1999 Elevated expression of the CC chemokine regulated on activation, normal T cell expressed and secreted (RANTES) in advanced breast carcinoma. Cancer Res 59:4681–4687

Maini RN, Taylor PC 2000 Anti-cytokine therapy for rheumatoid arthritis. Annu Rev Med 51:207–229

Moore RJ, Owens DM, Stamp G et al 1999 Mice deficient in tumour necrosis factor-α are resistant to skin carcinogenesis. Nat Med 5:828–831

Muller A, Homey B, Soto H et al 2001 Involvement of chemokine receptors in breast cancer metastasis. Nature 410:50–56

Negus RPM, Stamp GWH, Hadley J, Balkwill FR 1997 Quantitative assessment of the leukocyte infiltrate in ovarian cancer and its relationship to the expression of C-C chemokines. Am J Pathol 150:1723–1734

Nesbit M, Schaider H, Miller TH, Herlyn M 2001 Low-level monocyte chemoattractant protein-1 stimulation of monocytes leads to tumour formation in nontumorigenic melanoma cells. J Immunol 166:6483–6490

Orosz P, Kruger A, Hubbe M, Ruschoff J, Von Hoegen P, Mannel DN 1995 Promotion of experimental liver metastasis by tumor necrosis factor. Int J Cancer 60:867–871

Ramesh G, Reeves WB 2002 TNF-alpha mediates chemokine and cytokine expression and renal injury in cisplatin nephrotoxicity. J Clin Invest 110:835–842

Ristimäki R 2003 Cyclooxygenase 2: from inflammation to carcinogenesis. In: Cancer and inflammation. Wiley, Chichester (Novartis Found Symp 256) p 215–226

Scotton CJ, Wilson JL, Milliken D, Stamp G, Balkwill FR 2001 Epithelial cancer cell migration:a role for chemokine receptors? Cancer Res 61:4961–4965

Scotton CJ, Wilson JL, Scott K et al 2002 Multiple actions of the chemokine CXCL12 on epithelial tumor cells in human ovarian cancer. Cancer Res 62:5930–5938

Sica A, Saccani A, Bottazzi B et al 2000b Autocrine production of IL-10 mediates defective IL-12 production and NF-κ activation in tumor-associated macrophages. J Immunol 164:762–767

Tisdale MJ 2001 Cancer anorexia and cachexia. Nutrition 17:438–442

Tselepis C, Perry I, Dawson C et al 2002 Tumour necrosis factor-α in Barrett's oesophagus: a potential novel mechanism of action. Oncogene 21:6071–6081

Vicari AP, Caux C 2002 Chemokines in cancer. Cytokine Growth Factor Rev 13:143–154

Wiley HE, Gonzalez EB, Maki S, Wu M-T, Hwang ST 2001 Expression of CC chemokine receptor-7 and regional lymph node metastasis of B16 murine melanoma. J Natl Cancer Inst 93:1638–1643

Wilson J, Balkwill F 2002 The role of cytokines in the epithelial cancer microenvironment. Semin Cancer Biol 12:113–120

DISCUSSION

Wilkins: Presumably God or evolution, depending on your beliefs, gave mammals TNFα for some positive reason. Do your TNFα-negative mice show subtle immune system or developmental defects?

Balkwill: It is well documented that they have defects in B cells and regional lymph node development. They are susceptible to some intracellular pathogens. This is also seen in patients treated with TNF antagonists: if they have previously been infected with tuberculosis, the disease can be reactivated.

Wilkins: So there would be these side-effects to any therapeutic regime.

Balkwill: Yes, but the rheumatoid patients seem to do very well. I don't know how the companies got the courage to try anti-TNF in cancer: it was a big psychological hurdle to try to antagonize something called 'tumour necrosis factor'. I think one of the things that happened was that the anti-TNF antibody was being used at very high levels to treat graft versus host disease in cancer patients. This gave some confidence that it could be safely given to patients with cancer.

Gordon: Perhaps we should go back to calling TNF cachectin.

Balkwill: Certainly you could make a case for calling TNF TPF (tumour promoting factor), but this is a controversial area.

D'Incalci: Since you have lots of experience in ovarian cancer, and therapy of ovarian cancer is well established, at least for this class of compound, do you think there is some information about what happens in the tumour with respect to the macrophages and chemokines?

Balkwill: We discussed this earlier on: there is a suggestion that cisplatinum is toxic for tumour-associated macrophages. It is something that would be nice to study in more detail. It is possible that when there are so many macrophages in these tumours that the chemotherapy might well be knocking them out.

Gallimore: In the chemokine antagonist model, have you looked to see whether you need any immune cells for the tumour rejection? For example, if you deplete the CD8$^+$ T cells, does tumour rejection still occur?

Balkwill: It would be nice to do this sort of experiment. That's a good point.

Feldmann: You should be congratulated for your persistence in overcoming a Himalayan mountain of antagonism to TNF blockade in cancer. It is not just the psychological barrier, but also the commercial problem: once you start selling a drug, then the problem is that the company does not want to risk any side-effects in trials. People don't want to know in case it prejudices existing sales. But one of the things that might be worth considering if you have samples from the Enbrel-treated patients is to measure serum IL6 and CRP in the patients. One of the things that is quite unpredictable is how much inflammatory tissue you would need to raise these markers. They are raised in rheumatoid arthritis, they are not raised in MS, and surprisingly they are raised in acute coronary syndromes where, as far as we know, the inflammatory mass is pretty small. In some of these ovarian cancers it may be that you can get predictions of who is likely to do best with that therapy reasonably early, if they express markers of inflammation.

Balkwill: We have actually done that. We have studied MMP3, E selectin and soluble TNF receptor in Enbrel-treated cancer patients. There hasn't been any change in levels of these factors in the patients, but they did not have elevated levels of any of these markers to begin with

Feldmann: Did you look at VEGF?

Balkwill: No, because with VEGF there is a problem with the fact that platelets contain large amounts. In the trial we are just about to start we can study one aspect of the tumour microenvironment by sampling ovarian cancer ascites while the patients are being treated with Remicade.

Ristimäki: What about looking at peritoneal fluid?

Balkwill: That is what we will do in the next study.

Pepper: I was intrigued by your comments that your data are getting better with time. Have you moved to cleaner animal facilities?

Balkwill: No. It is a really practical thing. Our mice used to be 20 miles away. We used to grow the cells, harvest them and then send them 20 miles by taxi or van for injection into the mice. Now our mice are five minutes away and the tumours grow faster. The faster the tumour grows, the more effective the antagonism appears to be.

Pollard: Do you think the sole effect of the TNFα in the tumour is through its angiogenic properties, or does it work in other ways? If that is the case, for example, its role in killing things such as *Listeria* is rather different, can you go further downstream in its signalling pathways to pull out some interesting inhibitors of this specific pathway?

Balkwill: I can only really talk about ovarian cancer and the skin cancer models. In terms of TNF production, it is made by the epithelial cells in skin or the epithelial tumour cells. Epithelial cells are a major source of TNF, which is made in an autocrine fashion or in response to the chemokine CXCL12. I don't know whether it solely has an effect on angiogenesis. We are beginning to do animal tumour models with anti-TNF in some detail, so we hope to be able to dissect that out.

Pollard: You showed that MMP9 was up-regulated in the presence of TNFα, and Werb's idea is that MMP9 can activate EGF.

Balkwill: I think that MMP9 is also important. Primary keratinocytes from TNF knockout mice don't invade through matrigel. But because the effect of TNF on the skin carcinogenesis is so profound, I don't think there will be just one mechanism.

Blankenstein: You mentioned that TNF knockouts contained fewer initiated cells in the skin carcinogenesis model.

Balkwill: No, there are not fewer initiated cells. We looked at DNA adduct formation and the typical point mutation in c-Ha-ras. The level of both of these is identical in the knockout and wild-type mice. Our hypothesis is that the initiated cells don't get into the interfollicular epidermis: the cells are probably initiated in the bulge region of the hair follicle (infundibulum), but our hypothesis is that because they are not invasive they don't migrate to the interfollicular epidermis where they are in the right microenvironment for tumour promotion.

Ristimäki: To a certain extent this is in parallel with work published by Fürstenberger's group, who showed that under keratin 5 promoter *Cox2* transgene promotes tumorigenesis. In this classic skin cancer model that requires both initiation and promotion for tumour formation, in the case of *Cox2* transgene expression only initiation was needed for the tumours to occur. However, the

DNA adduct formation was the same with and without COX-2 (Müller-Decker et al 2002).

Pollard: We should have a complete renaming of all growth factors. Leukaemia inhibitory factor (LIF) has only ever inhibited one leukaemia, and Colin Stewart suggests it should be called BIF, for blastocyst implantation factor. We are trapped by the names.

Balkwill: It is a matter of balance. TNF is a licensed drug: in Europe it is licensed for treatment of irresectable soft tissue sarcoma by isolated limb perfusion in combination with melphalan and mild hypothermia. High doses of exogenous TNF promote tumour necrosis, mainly by destroying tumour blood vessels, whereas low doses of TNF produced endogenously in the tumour microenvironment may have tumour promoting activity.

Strieter: The definitions go back to the original experiments. 'Necrosis' came from the vascular component of TNF's effect for causing necrosis of tumours.

Reference

Müller-Decker K, Neufang G, Berger I, Neumann M, Marks F, Fürstenberger G 2002 Transgenic cyclooxygenase-2 overexpression sensitizes mouse skin for carcinogenesis. Proc Natl Acad Sci USA 99:12483–12488

In vivo manipulation of dendritic cell migration and activation to elicit antitumour immunity

Alain P. Vicari, Béatrice Vanbervliet, Catherine Massacrier, Claudia Chiodoni*, Céline Vaure, Smina Aït-Yahia, Christophe Dercamp, Fabien Matsos, Olivier Reynard, Catherine Taverne, Philippe Merle†, Mario P. Colombo*, Anne O'Garra‡[1], Giorgio Trinchieri and Christophe Caux[2]

*Schering-Plough Laboratory for Immunological Research, 69571 Dardilly, France, *Department of Experimental Oncology, Istituto Nazionale per lo Studio e la Cura dei Tumori, 20133 Milano, Italy, †INSERM U271, 69424 Lyon, France, and ‡DNAX Research Institute, Palo Alto, CA 94304, USA*

Abstract. Two approaches have been pursued to elicit antitumour immunity: (i) induce recruitment of immature dendritic cells or their precursors at a site of antigen delivery, and (ii) induce activation of tumour-infiltrating dendritic cells (DCs). The recruitment of selected DC subtype conditions the class of the immune response. Each immature DC population displays a unique spectrum of chemokine responsiveness. For examples, Langerhans cells (LCs) migrate selectively in response to CCL20/MIP-3α (through CCR6), blood CD11c+ DC to MCP chemokines (through CCR2). All these chemokines are inducible in response to inflammatory stimuli. CCL20/MIP-3α in particular is only detected within inflamed epithelium, at the site of antigen entry, which is infiltrated by immature DCs. Furthermore, to reach the site of injury, sequential responsiveness might operate, blood DC precursors are recruited by a set of chemokines (MIP, MCP) while within the tissue other chemokines will direct their navigation (CCL20/MIP-3α). Of interest, when injected *in vivo* together with antigen, MCP-4/CCL13, but not CCL20/MIP-3α, recruits blood monocytes or blood DC precursors that promptly differentiate into typical DCs and that improve antitumour immune responses. After antigen uptake, DCs acquire, upon maturation, responsiveness to CCR7 ligands (CCL21/SLC/6Ckine, CCL19/ELC/MIP-3β) due to receptor up-regulation. In particular, in the periphery, CCL21/SLC/6Ckine expressed by lymphatic vessels may direct into the lymph stream, antigen-loaded maturing DCs leaving the site of infection; while within lymph-node, CCL21/SLC/6Ckine plays a critical role in the entry of naïve T cells from the blood

---

[1]Present address: NIMR, The Ridgeway, Mill Hill, London, UK.
[2]This paper was presented at the symposium by Christophe Caux to whom all correspondence should be addressed.

through HEV. In regard to its central role, we decided to investigate whether the expression of CCL21/SLC/6Ckine in tumour may lead to antitumour immune responses. C26 colon carcinoma tumour cell line transduced with CCL21/SLC/6Ckine showed reduced tumorigenicity when injected *in vivo* into immunocompetent mice. The protection was CD8 dependent and associated with an important intratumoral infiltration of DCs. Most tumour infiltrating DCs (TIDCs) had an immature phenotype, were able to present TAA in the context of MHC class I, but were refractory to stimulation with the combination of LPS, IFNγ and anti-CD40 antibody. TIDC paralysis could be reverted, however, by *in vitro* or *in vivo* stimulation with the combination of a CpG immunostimulatory sequence and an anti-interleukin 10 receptor (IL10R) antibody. CpG or anti-IL10R alone were inactive in TIDC, while CpG triggered activation in normal DC. In particular, CpG plus anti-IL10R enhanced the TAA-specific immune response and triggered *de novo* IL-12 production. Subsequently, CpG plus anti-IL10R treatment showed robust antitumour therapeutic activity exceeding by far that of CpG alone, and elicited antitumour immune memory.

2004 Cancer and inflammation. Wiley, Chichester (Novartis Foundation Symposium 256) p 241–258

Life cycle of dendritic cells

Dendritic cells (DCs) are bone marrow-derived leukocytes which function as sentinels of the immune system (Banchereau & Steinman 1998, Sallusto & Lanzavecchia 1999, Steinman 1991). DC precursors migrate from the bone marrow through the bloodstream to almost every tissue, where they eventually become resident immature DCs. Langerhans cells (LCs) in the epidermis are the best studied example of immature DCs, which show a high ability for antigen uptake but a low capacity for antigen presentation. During pathogen invasion, immature DCs capture intruder antigens, and quickly leave the epidermis. They crawl through the dermis, cross the endothelium of lymphatic vessels and migrate to the draining lymph node. During their migration from the peripheral tissues, DCs undergo phenotypical and functional maturation. Most remarkably, they loose the ability to capture antigens and up-regulate the expression of co-stimulatory molecules. After reaching the subcapsular sinus of the lymph node, DCs move to the T cell areas through which T cells recruited from the blood percolate. T cell area DCs, known as interdigitating DCs (IDCs), are actively involved in the presentation of antigen to naïve T cells. The presentation of antigen to the appropriate T cells seems to be the ultimate mission of the DCs recruited from the periphery, as most of them die in the T cell areas, most likely by apoptosis. The complex pattern of DC migration favours the presentation of antigen captured at the periphery to the rare antigen-specific T cells, and the activation and subsequent clonal expansion of these T cells.

DCs perform different functions in distinct anatomical sites: steps of DC maturation

From the outlines of their life cycle, DCs appear as migratory cells moving from one site to the next to perform specific functions for which they acquire specific abilities through a stepwise maturation process. The analysis of *in vitro* generated DCs either from monocytes cultured in the presence of GM-CSF and interleukin (IL)4 (Sallusto & Lanzavecchia 1994) or from CD34$^+$ haematopoietic progenitor cells (HPCs) cultured in GM-CSF plus tumour necrosis factor (TNF)α (Caux et al 1992) has provided a model for the study of DC maturation and allowed the tentative identification of four major stages in the maturation process. (1) A heterogeneous stage of DC precursors represented *in vivo* by cells in the bone marrow and blood stream that are able to secrete pro-inflammatory and/or anti-viral cytokines. (2) A stage of immature DCs represented *in vivo* by cells resident within peripheral tissue (e.g. the skin LCs) that display very active receptor-dependent and independent endocytic activities but that do not express co-stimulatory molecules. (3) A stage of mature DCs, induced following exposure to pathogens (toll-like receptors) or pro-inflammatory mediators (IL1, TNFα), and able to efficiently prime naïve CD4$^+$ T cells. These mature DCs are represented *in vivo* by DCs migrating to lymph nodes and are characterized by the down-regulation of endocytic receptors and antigen uptake ability, translocation of the MHC class II molecules from the lysosomal compartment to the membrane (Cella et al 1997, de Saint-Vis et al 1998, Pierre et al 1997), and expression of co-stimulatory molecules. (4) A stage of mature/activated DCs induced by stimuli such as CD40L that induced the capacity to present antigen to naïve CD8$^+$ T cells and that represent the terminal differentiation of DCs (Albert et al 1998, Lanzavecchia 1998).

Sequential chemokine involvement to recruit immature DCs at sites of pathogen entry

Each step of DC trafficking involved in either their 'steady state' distribution in peripheral and lymphoid organs or in their recruitment upon inflammation/injury is likely to be controlled, at least partially, by soluble chemotactic factors known as chemokines (Dieu-Nosjean et al 1999). Different DC populations will likely not be recruited at the same anatomic site during different responses, and this will in part depend on the chemokine gradient released at the site of injury. It is expected that the type of resulting immune response will likely be dependent on the DC subpopulation recruited and its stage of maturation, and thus on the chemokines being secreted.

To reach the site of antigen deposition at epithelial surfaces, DCs have to pass the endothelial barrier, progress through the tissue (i.e. dermis) and cross the dermo-epithelial junction (basal membrane). In a recent study (Vanbervliet et al 2002) we observed that circulating blood DCs as well as monocytes express high levels of CCR2 and primarily respond to MCPs and not to MIP-3α/CCL20. In fact, we have not succeeded in the identification of circulating blood DCs or DC precursors expressing CCR6. Furthermore, while the CD34$^+$ HPC-derived CD1a$^+$ precursors committed to Langerhans cell differentiation primarily respond to MIP-3α/CCL20, the HPC-derived CD14$^+$ precursors respond to both MCPs and MIP-3α/CCL20. In concordance with the sequential expression of CCR2 and CCR6, the HPC-derived CD14$^+$ precursors initially acquire the ability to migrate in response to MCP-4/CCL13 and subsequently in response to MIP-3α/CCL20. Finally, we have observed that *in vivo* MIP-3α/CCL20 and MCP-4/CCL13 form complementary gradients in inflamed skin and mucosa. These observations suggest that the recruitment of DCs to the site of infection is controlled by the sequential action of different chemokines: (i) CCR2$^+$ circulating DCs or DC precursors are mobilized into the tissue, via the expression of MCPs by fibroblasts or endothelial cells and (ii) these cells traffic from the tissue to the site of pathogen invasion, via the production of MIP-3α/CCL20 by epithelial cells and the up-regulation of CCR6 in response to the tissue environment.

In vivo experiments in mouse tumour models have corroborated this hypothesis. Following intracutaneous injection, in contrast to MIP-3α/CCL20, MCP-4/CCL13 has the unique capacity to recruit DCs in the draining lymph node *in vivo*, and to increase antigen-specific immunity and antitumour immunity. The site of MCP-4/CCL13 injection is characterized by an influx of monocytic cells, although those lacked phenotypical features of DCs. When MCP-4/CCL13 injection was followed by β-galactosidase-encoding DNA plasmid vaccination, we observed an increase in antigen-specific serum IgG compared to plasmid alone. Moreover, MCP-4/CCL13 injection improved the effect of this vaccination scheme conferring protection against a C26 colon carcinoma tumour that expressed β-galactosidase as a model antigen. These results suggest that MCP-4/CCL13 is able to recruit blood monocytes or yet uncharacterized immature blood DC precursors that promptly differentiate into typical DCs, as previously reported in models of reverse endothelial transmigration (Randolph et al 1998a, 1999).

Recruitment of maturing DCs into draining node and initiation of the immune response

During injury, inflammatory mediators (e.g. IL1, TNF), infectious agent products (e.g. LPS, CpG, DNA) or T cell products (e.g. CD40L, IFNγ, IL17) will drive DC maturation and induce maturing DCs to emigrate out of the inflammatory site. A first consequence of this maturation will be the loss of functions characterizing immature DCs, in particular antigen uptake capacity as a consequence of cytoskeleton re-arrangement and endocytic receptor down-regulation. Furthermore, the responsiveness to most of the inflammatory chemokines (MIP-1α/CCL3, MIP-1β/CCL4, MCP1–4, RANTES/CCL5, MIP-3α/CCL20) is rapidly lost as a consequence of (i) a desensitization process involving saturation of their receptors by an endogenous production of ligands by activated DCs (Sallusto et al 1998, 1999) and (ii) a down-regulation of their receptor expression at the mRNA level (Dieu et al 1998, Foti et al 1999, Sallusto et al 1998, Sozzani et al 1998). Concomitantly to the loss of the inflammatory chemokine responsiveness, the CCR7 receptor, not expressed by immature DCs, is rapidly induced at the cell surface of maturing DCs.

The known ligands for CCR7 are 6Ckine/SLC/CCL21/Exodus-2 and ELC/MIP-3β/CCL19/Exodus-3 (Hedrick & Zlotnik 1997, Hromas et al 1997, Nagira et al 1997, Rossi et al 1997, Yoshida et al 1997). Contrary to many CC inflammatory chemokines ELC/MIP-3β/CCL19 and 6Ckine/SLC/CCL21 have no chemotactic activity on immature DCs. Both these ligands have been reported by several groups to mediate potent migration of human and mouse mature DCs through CCR7 (Chan et al 1999, Dieu et al 1998, Kellermann et al 1999, Saeki et al 1999, Sallusto et al 1998, Sozzani et al 1998, Yanagihara et al 1998) (Ogata et al 1999, Vecchi et al 1999). All human DC subsets (CD34+ progenitor-derived DCs, monocyte derived DCs, blood CD11c+, CD11c−) respond to ELC/MIP-3β/CCL19 and 6Ckine/SLC/CCL21 upon maturation (Dieu et al 1998, and our unpublished results).

Entry into lymphatic vessels, a control step?

In mouse, 6Ckine/SLC/CCL21 has been shown to be expressed by endothelial lymphatic vessels draining non-lymphoid tissues (Gunn et al 1998). Furthermore, exogenous 6Ckine/SLC/CCL21 has been shown to increase the yield of DCs emigrating out of mouse skin explants (Kellermann et al 1999), and blocking 6Ckine/SLC/CCL21 *in vivo* impaired emigration of DCs out of the dermis (Saeki et al 1999). Finally, we have found that, although 6Ckine/SLC/CCL21 is constitutively expressed on mouse endothelial lymphatics, its expression is strongly up-regulated a few hours following LPS injection (unpublished observation). This might suggest that the entry into lymphatics by maturing CCR7+ DCs is regulated by the level of 6Ckine/SLC/CCL21 expressed by lymphatic vessels, which is under the control of inflammatory stimuli. This possibility raises the question of the accessibility of draining lymphatics for *in vitro* generated DCs currently re-infused intra-dermally in DC-based antitumour immunotherapy clinical trials.

The notion that entry into lymphatics by maturing DCs could be a control step has already been suggested by the observation that antibody against the P glycoprotein (encoded by the multi-drug resistance gene, *MDR*) blocks the reverse migration of DCs, mimicking entry into lymphatics (Randolph et al 1998b). Although the role of P glycoprotein is not elucidated in this observation, it is suggested that the anti-MDR antibody blocks the release of certain eicosanoids such as LTC4 by activated DCs (Robbiani et al 2000). These molecules might deliver signal(s) to endothelial lymphatics for the control of vessel permeability.

Role of CCR7 ligands in the homing
of antigen loaded DCs in the T cell area

In mouse secondary lymphoid organs, 6Ckine/SLC/CCL21 is expressed on HEV (Gunn et al 1998). Similarly, in human, we observed 6Ckine/SLC/CCL21 expression in HEV as well as in numerous cells in the T cell-restricted areas (Dieu-Nosjean et al 1999). In human inflamed tonsils and lymph nodes, ELC/MIP-3β/CCL19 expression is also restricted to T cell areas, the homing site of mature IDCs. Similarly, in mice, ELC/MIP-3β/CCL19 is constitutively expressed within the T cell areas of different secondary lymphoid tissues (spleen, lymph nodes and Peyer's patches) (Ngo et al 1998).

Altogether, these observations suggest that, following inflammatory stimuli, DCs undergoing maturation express CCR7. Concomitantly, 6Ckine/SLC/CCL21 is induced on endothelial lymphatics during the inflammatory reaction and triggers the emigration of maturing DCs through the lymph stream. Mature DCs entering the draining lymph nodes are driven into the paracortical area in response to the production of ELC/MIP-3β/CCL19 and/or 6Ckine/SLC/CCL21 by cells spread over the T cell zone. The resistance of CCR7 to ligand-induced desensitization supports the hypothesis of a sequential role of 6Ckine/SLC/CCL21 and ELC/MIP-3β/CCL19 through CCR7 during migration of maturing DCs to the draining lymph node (Sallusto et al 1999).

The key roles of 6Ckine/SLC/CCL21 and ELC/MIP-3β/CCL19 in the recruitment of mature DCs into the T cell area of lymphoid organs has been recently demonstrated in naturally 6Ckine/SLC/CCL21 deficient mice (Gunn et al 1999) and in mouse deleted for CCR7 (Forster et al 1999). In these mice the anatomical structure of lymphoid organs is disorganized with a strong defect in naïve T cells homing in the T cell areas. In addition, in both strains, DCs fail to accumulate in the lymphoid organs following injection or contact sensitization. As a consequence, these animals have impaired immune responses with severely delayed antibody responses, lack of contact sensitivity and delayed-type hypersensitivity reactions, and a markedly increased susceptibility to infections.

Induction of antitumour immunity through SLC expression in tumours

In regard to its central role in regulating mature DCs and naïve T cells trafficking, we decided to investigate whether the expression of 6Ckine/SLC/CCL21 in tumours may lead to antitumour immune responses. C26 colon carcinoma tumour cell line transduced with 6Ckine/SLC/CCL21 showed reduced tumorigenicity when injected *in vivo*, in immunocompetent mice (Vicari et al 2000). The protection was CD8 dependent and associated with a heavy intratumoral infiltration of DCs. Surprisingly, although CCR7 mRNA expression was dramatically increased in 6Ckine/SLC/CCL21-transduced C26 tumours (C26-6CK), TIDCs resembled immature DCs. TIDC were CD11b$^+$, CD8α$^-$ and B220$^-$ in their vast majority, resembling the classical myeloid subset of DCs described in the mouse (Shortman & Liu 2002).

To test whether TIDCs from C26-6CK tumours had been able to capture and process tumour-associated antigens (TAAs) in the MHC class I pathway, we examined their capacity to present TAA-derived peptides to CTL, as previously described for a C26 tumour engineered to express GM-CSF and CD40L (Chiodoni et al 1999). The C26 (H-2^d) colon carcinoma expresses an immunodominant TAA which contains the L^d-restricted peptide AH-1 (Huang et al 1996), recognized by the E/88 CTL clone. The MC38 (H-2^b) colon carcinoma expresses the same TAA and contains a K^b-restricted peptide recognized by the TG905 CTL line (Sijts et al 1994). TIDCs purified from BALB/c x C57BL/6 (H-2^{dxb}) F1 mice bearing C26-6CK tumours were able to stimulate both CTL lines. In particular, the stimulation of the H-2^b TG905 CTL clearly indicates that TIDCs have been able to capture and present exogenous TAA in the MHC class I pathway.

Despite their ability to present tumour-derived antigens and induce CTL activity *in vivo*, these TIDCs were paralysed in an immature state. These observations prompted us to investigate chemokine-mediated DC recruitment together with activation signal(s) that would allow overcoming the TIDC paralysis.

Reversal of TIDC paralysis

Although the main function of DCs is probably to orchestrate a defence against pathogens, DCs are also well equipped to initiate antitumour responses (Banchereau & Steinman 1998). Indeed, *in vitro* generated DCs are able to sample and present tumour antigens for the priming of cytotoxic T cells (Ronchetti et al 1999) and DCs can produce the cytokines IL12, TNFα and IFNα that play diverse roles in antitumour immune responses (Banchereau & Steinman 1998). Accordingly, several investigators have successfully harnessed this biological potential of DCs by preventing or curing transplantable tumours in mice following the infusion of *ex vivo* derived DCs pulsed with TAAs (Mayordomo et al 1995, Paglia et al 1996). Today, this strategy is being evaluated in clinical trials investigating what is the most efficient subset of the DC form of antigen and activation stimulus (Dallal & Lotze 2000).

Although solid human tumours are frequently infiltrated by DCs (Bell et al 1999, Enk et al 1997, Scarpino et al 2000), only a few studies have focused on the therapeutic potential of TIDCs. Yet, in a mouse model, TIDC were shown to have captured and processed TAAs exogenously in the MHC class I pathway (Chiodoni et al 1999), suggesting that this first step towards immune reactivity was not impaired. However, the tumour milieu appears to lack the expression of DC activation factors such as microbial stimuli which are known to be crucial in DC physiology (Reis e Sousa 2001). Furthermore, tumour cells or tumour-infiltrating cells may produce IL10, prostaglandin E2 (PGE$_2$) and TGFβ that impair DC functions (Chouaib et al 1997, Vicari et al 2002a). Thus, lack of activation in conjunction with inhibition of DCs within tumours could explain why the immune response against the tumour is not taking place. Therefore, manipulations aimed at restoring TIDC function may provide novel immunotherapeutic strategies against cancer.

Combination of CpG and anti-IL10R antibody overcomes TIDC paralysis in vitro

We isolated TIDCs from various transplantable tumours as well as from hepatocarcinoma developing in X/*myc* transgenic mice (Terradillos et al 1997). TIDCs isolated from C26-6CK were identical to TIDCs isolated from parental C26 tumours or from other transplantable tumours, thus resembling the classical myeloid subset of DC (CD11b$^+$, CD8α^- and B220$^-$). On the other hand, TIDCs isolated from liver hepatocarcinoma were more diverse, including CD11b$^+$ and CD11b$^-$ DCs as well as cells expressing CD8α and/or B220, the latter marker being ascribed to mouse type I interferon-producing cells (Asselin-Paturel et al 2001, Shortman & Liu 2002). All these TIDC subsets had an immature phenotype, with intermediate levels of surface MHC class II and no detectable CD40 or CD86 molecules, with the exception of the B16 melanoma TIDCs which expressed low levels of CD40 and CD86. A feature of immature DCs is a response to stimulation with LPS + IFNγ + anti-CD40 antibody by increasing the expression of CD40 and CD86 as well as by producing IL12 p70. In contrast, TIDCs from C26-6CK tumours maintained an immature phenotype under activation and did not produce detectable IL12 p70. Similarly, DCs from normal liver produced IL12 p70 in response to LPS + IFNγ + anti-CD40 while TIDCs from hepatocarcinoma did not. Last, a supernatant from C26 tumours added at the time of activation abolished the secretion of IL12 p70 in bone marrow DCs.

These results indicate that TIDCs were refractory to LPS + IFNγ + anti-CD40 stimulation.

We tested different combinations of substances with the aim of relieving tumour-mediated inhibition and simultaneously mediating DC activation, including combinations of the TLR-9 ligand CpG 1668 (Hemmi et al 2000) and an anti-IL10R blocking antibody (Castro et al 2000). In control bone marrow DCs, CpG alone was able to induce the secretion of IL12 p70 and TNFα in amounts similar to that obtained with LPS + IFNγ + anti-CD40. The addition of anti-IL10R or anti-IL10 antibody to BM-DC culture increased the production of IL12 p70 by about 30%. Anti-IL10R alone or CpG alone were not able to restore significant IL12 p70 production by TIDCs cultured in the presence of LPS + IFNγ + anti-CD40. In marked contrast, the combination of CpG and anti-IL10R antibody induced the secretion by TIDCs of large quantities of IL12 p70 as well as TNFα and was also the best activation condition for mixed lymphocyte reaction (MLR).

When extended to other transplantable tumour (C26, B16F0, LL2 and TSA) as well as from X/*myc* hepatocarcinoma, TIDCs isolated from these tumours did not respond to LPS + IFNγ + anti-CD40 but showed a robust response to CpG + anti-IL10R for IL12 p70 production. Thus, we identified the combination of CpG plus anti-IL10R antibody as a novel and unique way to activate TIDCs from tumours from different histological origin as well as in a transgenic tumour model (Vicari et al 2002b).

Combination of CpG and anti-IL10R antibody
has therapeutic antitumour effect in vivo

When C26-6CK tumour-bearing mice were injected with CpG intratumourally and/or with anti-IL10R antibody intraperitoneally, we observed that CpG + anti-IL10R induced the secretion of intracellular IL-12 p40/70 in a large proportion of TIDCs as soon as 2 hours after treatment, while CpG and anti-IL10R alone had no effect.

We treated mice implanted with subcutaneous C26 or B16F0 tumours with various combinations of CpG 1668, control GL113 antibody or anti-IL10R antibody. We observed that anti-IL10R or CpG alone had no effect on tumour incidence or survival in the C26 model, nor in the B16F0 model. In marked contrast, combination of CpG plus anti-IL10R had a significant effect on tumour incidence and on survival in both models.

In vivo experiments using depleting antibodies suggest that several components of the adaptive and innate immune response, including CD4⁺, CD8⁺ T cells and NK cells, might contribute to tumour eradication. When C26 tumour-bearing SCID mice were injected with CpG plus anti-IL10R, no significant impact of the treatment on tumour incidence and survival was observed, strongly suggesting a role for T cells in CpG + anti-IL10R treatment in immunocompetent mice.

We then analysed the capacity of TIDCs from C26-6CK tumours to induce TAA-specific responses *in vivo*. Following enrichment and overnight activation, we injected TIDCs into naïve mice and five days later we measured the AH1-specific response. We found that TIDCs from C26-6CK tumours were able to induce AH-1-specific cytotoxicity, as reported previously for C26-GM-CD40L tumours (Chiodoni et al 1999). TIDCs activated with CpG plus anti-IL10R were able to increase the number of AH-1-specific IFNγ-producing cells. Collectively, these results strongly suggest that activation with CpG plus anti-IL10R antibody increases TAA-specific immune responses.

Last, we analysed the antitumour immune memory response in mice cured of C26 tumours with CpG + anti-IL10R treatment by rechallenging the animals with C26 cells, 45 days after the first challenge. We observed that only 25% of the mice developed tumours, and with a marked delay when compared to naïve mice, suggesting the establishment of an antitumour immune memory response following CpG + anti-IL10R therapy (Vicari et al 2002b).

Conclusion

DCs induce, sustain and regulate immune responses. The different classes of immune responses are likely the consequence of the combination of (i) the presence of different DC subsets with specific biological functions, such as polarization of T cell responses towards type 1 or type 2, regulation of B cell responses, or induction of antiviral immunity. (ii) The maturation stage of the DCs will also determine the outcome of the immune response: at the precursor stage, DCs have the ability to secrete large amounts of pro-inflammatory and/or antiviral cytokines; at the immature stage they display high antigen uptake capacity; and at the mature stage they are able to activate and modulate T cell responses. (iii) Finally, the type of maturation signals, the microenvironment, and the type and dose of antigen, will modulate the type of DC maturation and consequently the class of the immune responses.

The control of DC trafficking appears to be a complex process with the intervention of several chemokines and many other molecules such as selectins, integrins, and proteases. Future investigations will be required to assess the role of chemokines in constitutive DC trafficking versus induced DC mobilization during inflammation. Such approaches might help in understanding the role of DCs in peripheral tolerance maintenance versus immune induction.

Finally, understanding pathogen/DC interactions should allow identifying pathways for DC activation that should lead to avenues of immune intervention against tumours. However, using a microbial stimulus by itself may not be effective in some cancers, if DCs become refractory because of tumour-derived factors. Using antagonists of tumour inhibitory factors such as anti-IL10R may significantly lower the threshold required for TIDC activation to a point amenable for successful immunotherapy using biological modifiers such as CpG sequences.

Acknowledgements

We thank C. Alexandre, M. Vatan and D. Lepot for editorial assistance; Doctors and colleagues from clinics and hospitals in Lyon who provide us with umbilical cord blood samples and tonsils.

References

Albert ML, Sauter B, Bhardwaj N 1998 Dendritic cells acquire antigen from apoptotic cells and induce class I-restricted CTLs. Nature 392:86–89

Asselin-Paturel C, Boonstra A, Dalod M et al 2001 Mouse type I IFN-producing cells are immature APCs with plasmacytoid morphology. Nat Immunol 2:1144–1150

Banchereau J, Steinman RM 1998 Dendritic cells and the control of immunity. Nature 392:245–252

Bell D, Chomarat P, Broyles D et al 1999 In breast carcinoma tissue, immature dendritic cells reside with the tumor while mature dendritic cells are located in peritumoral areas. J Exp Med 190:1417–1426

Castro AG, Neighbors M, Hurst SD et al 2000 Anti-interleukin 10 receptor monoclonal antibody is an adjuvant for T helper cell type 1 responses to soluble antigen only in the presence of lipopolysaccharide. J Exp Med 192:1529–1534

Caux C, Dezutter-Dambuyant C, Schmitt D, Banchereau J 1992 GM-CSF and TNF-alpha cooperate in the generation of dendritic Langerhans cells. Nature 360:258–261

Cella M, Engering A, Pinet V, Pieters J, Lanzavecchia A 1997 Inflammatory stimuli induce accumulation of MHC class II complexes on dendritic cells. Nature 388:782–787

Chan VW, Kothakota S, Rohan MC et al 1999 Secondary lymphoid-tissue chemokine (SLC) is chemotactic for mature dendritic cells. Blood 93:3610–3616

Chiodoni C, Paglia P, Stoppacciaro A, Rodolfo M, Parenza M, Colombo MP 1999 Dendritic cells infiltrating tumors cotransduced with granulocyte/macrophage colony-stimulating factor (GM-CSF) and CD40 ligand genes take up and present endogenous tumor-associated antigens, and prime naive mice for a cytotoxic T lymphocyte response. J Exp Med 190:125–133

Chouaib S, Asselin-Paturel C, Mami-Chouaib F, Caignard A, Blay JY 1997 The host-tumor immune conflict: from immunosuppression to resistance and destruction. Immunol Today 18:493–497

Dallal RM, Lotze MT 2000 The dendritic cell and human cancer vaccines. Curr Opin Immunol 12:583–588

de Saint-Vis B, Vincent J, Vandenabeele S et al 1998 A novel lysosome associated membrane glycoprotein, DC-LAMP, induced upon DC maturation, is transiently expressed in MHC class II compartment. Immunity 9:325–336

Dieu MC, Vanbervliet B, Vicari A et al 1998 Selective recruitment of immature and mature dendritic cells by distinct chemokines expressed in different anatomic sites. J Exp Med 188:373–386

Dieu-Nosjean MC, Vicari A, Lebecque S, Caux C 1999 Regulation of dendritic cell trafficking: a process which involves the participation of selective chemokines. J Leukoc Biol 66:252–262

Enk AH, Jonuleit H, Saloga J, Knop J 1997 Dendritic cells as mediators of tumor-induced tolerance in metastatic melanoma. Int J Cancer 73:309–316

Forster R, Schubel A, Breitfeld D et al 1999 CCR7 coordinates the primary immune response by establishing functional microenvironments in secondary lymphoid organs. Cell 99:23–33

Foti M, Granucci F, Aggujaro D et al 1999 Upon dendritic cell (DC) activation chemokines and chemokine receptor expression are rapidly regulated for recruitment and maintenance of DC at the inflammatory site. Int Immunol 11:979–986

Gunn MD, Tangemann K, Tam C, Cyster JG, Rosen SD, Williams LT 1998 A chemokine expressed in lymphoid high endothelial venules promotes the adhesion and chemotaxis of naive T lymphocytes. Proc Natl Acad Sci USA 95:258–263

Gunn MD, Kyuwa S, Tam C et al 1999 Mice lacking expression of secondary lymphoid organ chemokine have defects in lymphocyte homing and dendritic cell localization. J Exp Med 189:451–460

Hedrick JA, Zlotnik A 1997 Identification and characterization of a novel beta chemokine containing six conserved cysteines. J Immunol 159:1589–1593

Hemmi H, Takeuchi O, Kawai T et al 2000 A Toll-like receptor recognizes bacterial DNA. Nature 408:740–745

Hromas R, Gray PW, Chantry D et al 1997 Cloning and characterization of exodus, a novel beta-chemokine. Blood 89:3315–3322

Huang AY, Gulden PH, Woods AS et al 1996 The immunodominant major histocompatibility complex class I-restricted antigen of a murine colon tumor derives from an endogenous retroviral gene product. Proc Natl Acad Sci USA 93:9730–9735

Kellermann SA, Hudak S, Oldham ER, Liu YJ, McEvoy LM 1999 The CC chemokine receptor-7 ligands 6Ckine and macrophage inflammatory protein-3 beta are potent chemoattractants for in vitro- and in vivo-derived dendritic cells. J Immunol 162:3859–3864

Lanzavecchia A 1998 Immunology. Licence to kill. Nature 393:413–414

Mayordomo JI, Zorina T, Storkus WJ et al 1995 Bone marrow-derived dendritic cells pulsed with synthetic tumour peptides elicit protective and therapeutic antitumour immunity. Nat Med 1:1297–1302

Nagira M, Imai T, Hieshima K et al 1997 Molecular cloning of a novel human CC chemokine secondary lymphoid- tissue chemokine that is a potent chemoattractant for lymphocytes and mapped to chromosome 9p13. J Biol Chem 272:19518–19524

Ngo VN, Tang HL, Cyster JG 1998 Epstein-Barr virus-induced molecule 1 ligand chemokine is expressed by dendritic cells in lymphoid tissues and strongly attracts naive T cells and activated B cells. J Exp Med 188:181–191

Ogata M, Zhang Y, Wang Y et al 1999 Chemotactic response toward chemokines and its regulation by transforming growth factor-beta1 of murine bone marrow hematopoietic progenitor cell-derived different subset of dendritic cells. Blood 93:3225–3232

Paglia P, Chiodoni C, Rodolfo M, Colombo MP 1996 Murine dendritic cells loaded in vitro with soluble protein prime cytotoxic T lymphocytes against tumor antigen in vivo. J Exp Med 183:317–322

Pierre P, Turley SJ, Gatti E et al 1997 Developmental regulation of MHC class II transport in mouse dendritic cells. Nature 388:787–792

Randolph GJ, Beaulieu S, Lebecque S, Steinman RM, Muller W 1998a Differentiation of monocytes into dendritic cells in a model of transendothelial trafficking. Science 282:480–483

Randolph GJ, Beaulieu S, Pope M et al 1998b A physiologic function for p-glycoprotein (MDR-1) during the migration of dendritic cells from skin via afferent lymphatic vessels. Proc Natl Acad Sci USA 95:6924–6929

Randolph GJ, Inaba K, Robbiani DF, Steinman RM, Muller WA 1999 Differentiation of phagocytic monocytes into lymph node dendritic cells in vivo. Immunity 11:753–761

Reis e Sousa C 2001 Dendritic cells as sensors of infection. Immunity 14:495–498

Robbiani DF, Finch RA, Jager D, Muller WA, Sartorelli AC, Randolph GJ 2000 The leukotriene C(4) transporter MRP1 regulates CCL19 (MIP-3beta, ELC)-dependent mobilization of dendritic cells to lymph nodes. Cell 103:757–768

Ronchetti A, Rovere P, Iezzi G et al 1999 Immunogenicity of apoptotic cells in vivo: role of antigen load, antigen-presenting cells, and cytokines. J Immunol 163:130–136

Rossi DL, Vicari AP, Franz-Bacon K, McClanahan TK, Zlotnik A 1997 Identification through bioinformatics of two new macrophage proinflammatory human chemokines: MIP-3alpha and MIP-3beta. J Immunol 158:1033–1036

Saeki H, Moore AM, Brown MJ, Hwang ST 1999 Cutting edge: secondary lymphoid-tissue chemokine (SLC) and CC chemokine receptor 7 (CCR7) participate in the emigration pathway of mature dendritic cells from the skin to regional lymph nodes. J Immunol 162:2472–2475

Sallusto F, Lanzavecchia A 1994 Efficient presentation of soluble antigen by cultured human dendritic cells is maintained by granulocyte/macrophage colony-stimulating factor plus interleukin 4 and downregulated by tumor necrosis factor alpha. J Exp Med 179:1109–1118

Sallusto F, Lanzavecchia A 1999 Mobilizing dendritic cells for tolerance, priming, and chronic inflammation. J Exp Med 189:611–614

Sallusto F, Schaerli P, Loetscher P et al 1998 Rapid and coordinated switch in chemokine receptor expression during dendritic cell maturation. Eur J Immunol 28:2760–2769

Sallusto F, Palermo B, Lenig D et al 1999 Distinct patterns and kinetics of chemokine production regulated dendritic cell function. Eur J Immunol 29:1617–1625

Scarpino S, Stoppacciaro A, Ballerini F et al 2000 Papillary carcinoma of the thyroid: hepatocyte growth factor (HGF) stimulates tumor cells to release chemokines active in recruiting dendritic cells. Am J Pathol 156:831–837

Shortman K, Liu YJ 2002 Mouse and human dendritic cell subtypes. Nat Rev Immunol 2:151–161

Sijts AJ, Ossendorp F, Mengede EA, van den Elsen PJ, Melief CJ 1994 Immunodominant mink cell focus-inducing murine leukemia virus (MuLV)-encoded CTL epitope, identified by its MHC class I-binding motif, explains MuLV-type specificity of MCF-directed cytotoxic T lymphocytes. J Immunol 152:106–116

Sozzani S, Allavena P, D'Amico G et al 1998 Differential regulation of chemokine receptors during dendritic cell maturation: a model for their trafficking properties. J Immunol 161:1083–1086

Steinman RM 1991 The dendritic cell system and its role in immunogenicity. Annu Rev Immunol 9:271–296

Terradillos O, Billet O, Renard CA et al 1997 The hepatitis B virus X gene potentiates c-myc-induced liver oncogenesis in transgenic mice. Oncogene 14:395–404

Vanbervliet B, Homey B, Durand I et al 2002 Sequential involvement of CCR2- and CCR6-ligands for immature dendritic cell recruitment: possible role at inflamed epithelial surfaces. Eur J Immunol 32:231–242

Vecchi A, Massimiliano L, Ramponi S et al 1999 Differential responsiveness to constitutive vs. inducible chemokines of immature and mature mouse dendritic cells. J Leukoc Biol 66:489–494

Vicari A, Ait-Yahia S, Chemin K, Mueller A, Zlotnik A, Caux C 2000 Antitumor effects of the mouse chemokine 6Ckine/SLC through angiostatic and immunological mechanisms. J Immunol 165:1992–2000

Vicari A, Caux C, Trinchieri G 2002a Tumour escape from immune surveillance through dendritic cell inactivation. Semin Cancer Biol 12:33–42

Vicari AP, Chiodoni C, Vaure C et al 2002b Reversal of tumor-induced dendritic cell paralysis by CpG immunostimulatory oligonucleotide and anti-interleukin 10 receptor antibody. J Exp Med 196:541–549

Yanagihara S, Komura E, Nagafune J, Watarai H, Yamaguchi Y 1998 EBI1/CCR7 is a new
    member of dendritic cell chemokine receptor that is up-regulated upon maturation. J
    Immunol 161:3096–3102
Yoshida R, Imai T, Hieshima K et al 1997 Molecular cloning of a novel human CC chemokine
    EBI1-ligand chemokine that is a specific functional ligand for EBI1, CCR7. J Biol Chem
    272:13803–13809

DISCUSSION

Pollard: One thing with human tumours, especially those with very high levels of CSF1, is that CSF1 tends to immunosuppress through inhibition of the maturation of DCs. This might be something you could manipulate to try to activate DCs locally.

Caux: There is a clear effect of CSF1 in combination with IL6 in blocking DC differentiation. This has been reported in several tumour models. It might be too late to block CSF1 at the same time as the activation. The effect of CSF1 is to block differentiation of DCs to favour macrophages versus DCs. If you block CSF1 at the time you activate, it may not work: you need to block it before.

Strieter: There is one other caveat in terms of mouse versus human tumours. 6Ckine/SLC/CCL21 is a unique CC chemokine that in the mouse uses two different receptors, CXCR3 and CCR7. In contrast, in humans it only uses CCR7. This somewhat complicates the complexity of some of your results in mouse as opposed to what you might expect to see in human tumours.

Caux: What we showed with the mouse 6Ckine/SLC was the same with the human 6Ckine/SLC, which does not bind to the mouse CXCR3. The same observation is made in terms of the immune response. However, there is probably an effect on angiogenesis, which I didn't show. There is a five day delay of tumour growth with mouse 6Ckine/SLC which is not apparent with the human 6Ckine/SLC.

Brennan: Is IL10 the only factor in the tumour cell supernatant that modulates the DC function?

Caux: Probably not. One thing I didn't mention is that this is an effect of the tumour supernatant on DCs that we are describing. It is not a supernatant from the cell lines. IL10 does not come from the tumour itself, but it can be produced by the DCs themselves. There are other factors produced in the tumour environment such as PGE_2 that induce IL10. In this model, a colon carcinoma model, this is probably a prominent factor to induce IL10.

Brennan: At what level is the inability of the DCs to respond to LPS or CD40 mediated? Is it TLR dependent?

Caux: There is no good reagent for looking at the cell surface. On the basis of PCR it doesn't seem that there is a deficiency in terms of expression of TLR4.

Cerundolo: You mentioned that the regression of the tumours is dependent on CD4$^+$CD8$^+$ cells. Have you looked at expansion of any antigen-specific T cell responses?

Caux: This has been done by Elispots. There is induction of specific immune responses, but also some non-specific responses. It has not been examined by tetramers in this model.

Cerundolo: Is there any class I restricted epitope?

Caux: There is, the C26 (H-2^d) colon carcinoma expresses an immuno-dominant TAA which contains the L^d-restricted peptide AH-1, recognized by the E/88 CTL clone. We analysed the capacity of TIDCs from C26-6CK tumors to induce TAA-specific responses *in vivo*. Following enrichment and overnight activation, we injected TIDCs into naïve mice and five days later we measured the AH-1-specific response. We found that TIDCs from C26-6CK tumors were able to induce AH-1-specific cytotoxicity. TIDC activated with CpG plus anti-IL10R were able to increase the number of AH-1-specific IFNγ-producing cells.

Pepper: You said that 6Ckine drives DCs into lymphatics. What is the evidence for this, and what is the mechanism?

Caux: In the CCR7 knockout there are fewer cells that emigrate from the skin. There is a strong suggestion in the literature that 6Ckine plays a role at the level of the entry into lymphatics. Although it has been proposed to be constitutively expressed, it can also be up-regulated upon inflammation. When we look in mouse after LPS injection, which drives DC emigration out of the skin, there is up-regulation of 6Ckine by lymphatics (A. P. Vicari, unpublished results). This is probably the first step that regulates the immigration of DCs.

Gallimore: Have you looked at whether the DCs that you isolate from the tumours can somehow activate regulatory T cells? Could you do some sort of *in vitro* assay to test this?

Caux: That is an important question that has not been addressed. The problem for the regulatory T cell is what is the read out? Suppression of naïve T cell activation?

Gallimore: You could compare the ability of DCs isolated from tumours and other DCs to stimulate T cell responses both in the presence and absence of regulatory cells. This might give some indication of whether DCs isolated from tumours activate regulatory cells better than the normal DCs.

Caux: We are starting to do this type of experiment to see whether TIDCs can activate regulatory T cells or induce their development from naïve T cells.

Blankenstein: A couple of years ago we did some experiments with DC vaccines using β galactosidase as a model antigen and analysed tumour rejection in F1 mice. At that time we struggled with the problem of hybrid cell resistance. Do you know whether your cross-presentation could be facilitated by hybrid cell resistance?

Caux: I'm not sure I understand the definitions.

Blankenstein: Hybrid cell resistance means that cells, in particular haematopoietic cells with the haplotype of one parental strain are rejected in F1 animals. Probably, this is due to missing H2 in certain combinations which leads to NK-mediated rejection.

Caux: The experiment that was done there was not *in vivo*: the DCs were isolated from tumours implanted into F1 mice and put back *in vitro* with T cell clones. This was not a primary immune response we were looking at in these F1 mice. We just demonstrated that TIDCs isolated from tumour can present the TAA both in H2B and H2D, showing that they have captured antigen from the tumour.

Oppenheim: Most of the DCs you are talking about are myeloid. In the tumour you looked at, did you also detect the plasmacytoid DCs?

Caux: We are looking at these cells in mouse tumour models. So far, in the transplantable tumour models, we have mainly observed CD11c$^+$, CD11b$^+$, CD8$^-$, B220$^-$, corresponding to 'myeloid' DCs in mice. In the hepatocarcinoma model there seems to be more heterogeneity, so this can be a more interesting model in terms of plasmacytoid DCs. The plasmacytoid DCs are the IFNα-producing cells in the blood in response to viruses.

Oppenheim: The plasmacytoid DCs are supposed to be better responders to the CpG by the TLR9.

Caux: There is a big difference in mouse. In mouse TLR9 is expressed both in myeloid and plasmacytoid DCs, and clearly the effect we see there is probably via the myeloid DCs. If we deplete the plasmacytoid DCs there is no effect. The main difficulty for extending these observations to human is that in humans the TLR9 is only expressed on plasmacytoid DCs and not on 'myeloid' DCs.

Oppenheim: It was my understanding from the DC meeting that the human myeloid DCs still respond to CpG although people aren't clear about how this happens. The conclusion was that the cells show plasticity. Somehow the human myeloid DCs also respond to CpG even though the receptor story is not clear.

Caux: There is probably cross-talk between plasmacytoid DCs and 'myeloid' DCs. If you put CpG in a mixture of DCs, then there is activation of myeloid DCs, but this is probably an indirect effect.

Oppenheim: Are human tumours infiltrated by both types of DC?

Caux: This isn't really known. There is a lot of literature showing presence of 'myeloid' DCs in many types of tumours. There is one publication in ovarian tumours showing presence of pDC in ascites (Zou et al 2001).

Cerundolo: We have been looking at primary melanomas in patients, specifically looking for plasmacytoid DCs in infiltrating primary tumours. We can see them, but they appear to have an immature phenotype.

Feldmann: Where does the IL10 come from in your experiments?

Caux: *In vitro* it clearly comes from DCs. *In vivo* I don't know: it probably comes from a number of sources, including macrophages and T cells in the tumour.

Feldmann: Is there a literature about tumour cells making IL10?

Caux: Yes there is lots of information on this that has been published.

Strieter: We published several years ago that non-small-cell lung cancer cells can make IL10.

Balkwill: Ovarian cancer cells do, too.

Gordon: You seem more optimistic about these *in vivo* protocols for using DCs in melanoma than some of the other people I have heard talk about this. What are the grounds for your optimism? You said 20–30% of them showed some benefit.

Caux: I have in mind the clinical trial that Jacques Banchereau carried out (Banchereau et al 2001).

Gordon: He is always optimistic!

Caux: The difficulty with these clinical trials is the late-stage of the melanoma patients. However, there is a clear effect there.

Cerundolo: In future clinical trials it will be important to ensure that clinical efficacy is not the only end point of the trial, since very often there is no correlation between partial or complete tumour regressions and expansion of tumour specific immune responses, as measured by Elispot and tetramer. In the clinical trial that Banchereau published, the readout was Elispots. Although he didn't show any background — the background was subtracted — the impression was that there was induction of antigen-specific T cells. I have been directly involved in the clinical trial that Gerold Schuler has done by analysing some patients' samples. I can confirm that in his trial we did see some expansion of antigen-specific T cells. This result has been confirmed by Thierry Boon looking at MAGE1/A1-specific CTL. But this expansion is not huge, but it is definitely above the background.

Blankenstein: Perhaps one should add that in the many melanoma vaccine trials, one often finds around a 10% response, similar to what was found long ago when BCG was injected into melanoma lesions. This might not necessarily correlate with CTL responses, as Enzo Cerundolo pointed out. Do you think that a DC vaccine means a qualitatively different vaccine compared to all the other vaccines we have, for example such adjuvants that directly target the APC *in vivo*?

Caux: I started my presentation with the clinical trial using infusion of antigen-loaded DCs because it demonstrated that DCs can stimulate antitumour immune responses in human patients. I am not saying that is the only way to do it. This is part of a rationale for us to try to define ways to activate DCs *in vivo*. BCG was probably doing what I was just showing: to activate the DCs in the tumour through a toll-like receptor.

References

Banchereau J, Palucka AK, Dhodapkar M et al 2001 Immune and clinical responses in patients with metastatic melanoma to CD34(+) progenitor-derived dendritic cell vaccine. Cancer Res 61:6451–6458

Zou W, Machelon V, Coulomb-L'Hermin A et al 2001 Stromal-derived factor-1 in human tumors recruits and alters the function of plasmacytoid precursor dendritic cells. Nat Med 12:1339–1346

Final general discussion

D'Incalci: One aspect that we haven't discussed much relates to the possibility that inflammation can have an effect on the current therapies that are performed. I have an interesting story related to Glivec, which is an inhibitor of tyrosine kinase developed by Novartis. It was selected for its high specificity for Abelson tyrosine kinase and c-kit. This drug has shown activity against chronic myeloid leukaemia and gastrointestinal sarcomas involving c-kit. In collaboration with Gambacorti Passerini at the National Cancer Institute of Milan we investigated a human chronic myeloid leukaemia growing in mice that became resistant to Glivec. In most cases the resistance to this drug is due to mutation of the target protein, whereas in our model the resistance was associated with progression of the tumour. If the tumour was implanted and immediately treated the animals were cured. If, however, the tumours were allowed to grow and were then treated the mice were not cured. These cells were just as sensitive to Glivec as they were before, because if we transplanted them to a new animal and treated straight away, the cure was observed. There is therefore some host factor present when the tumour was advanced that was inactivating the drug. We found that this factor is α1-acid glycoprotein, which is an acute-phase-reaction protein made by the liver. This protein binds with high affinity to Glivec, reducing the penetration of the drug into cells. In this experimental model we could counteract this resistance mechanism by displacing the binding with erythromycin. These results have been obtained in mice and we don't know whether this finding is clinically relevant. Nevertheless a large body of data exists suggesting that α1-acid glycoprotein can increase in cancer patients. There are old literature reports on the determination of serum levels of α1-acid glycoprotein as a prognostic marker for ovarian cancers. Certainly α1-acid glycoprotein can be very variable in cancer patients. This is just an example showing that we know very little about this kind of phenomenon. The potential relevance of the production of proteins that can interact with drugs should be investigated.

I was very interested in the potential role of chemokines or other inflammation related factors that can regulate survival and apoptosis. The propensity of cells to activate apoptosis is crucial for the efficacy of anticancer drugs. Whether host factors can modulate this phenomenon is unknown and merits exploration. Another issue related to the influence of inflammation on the efficacy of drug treatment concerns metabolism. There is some evidence that cytokines can modify some p450 isoenzymes. I think that there is a broad use of corticosteroids associated with several cancer therapies. Since corticosteroids are well known modifiers of metabolism their use could alter the activity of many chemotherapeutic drugs. We also didn't discuss whether and to what extent anticancer drugs modify host responses, for instance with respect to angiogenesis. There are already clinical trials looking at the effects of low doses of antimitotic agents that according to some experimental studies should inhibit angiogenesis. This is an aspect that should be evaluated in more detail. Perhaps for many drugs in clinical development it will be important from the beginning to plan the evaluation of these aspects.

There are several signal translation inhibitors that are under clinical development as potential novel anticancer drugs. Clinical and laboratory attention is focused on the effects of these drugs on the tumour cells. However these compounds can certainly exhibit profound effects on the regulation of the immune system. If we take as an example the inhibitors of PI3 kinases or the inhibitors of the proteosome, these would be expected to show a strong effect on a variety of systems. Drug developers should collaborate with immunologists to evaluate these new drugs. It is very important not only to interpret the results (i.e. antitumour effect, toxicity), but also to develop intelligent strategies for a rational combination of different therapeutic approaches.

Strieter: I believe there is actually a trial looking at inhaled steroids and airway cell dysplasia.

Mantovani: I wanted to continue on the subject of histone deacetylase. There is a recent paper showing that a histone deacetylase inhibitor is a very strong inhibitor of inflammatory cytokine production of TNF and IL1 (Leoni et al 2002). Have people looked at this in the anticancer clinical trials?

D'Incalci: I don't know, but I suspect not. I don't think that in many trials these kinds of aspects are really considered.

Balkwill: You could argue that it doesn't matter: if they work, who cares what they do *in vivo*? If we do care how they work *in vivo*, how would we study this? How would we get any idea whether your signal transduction inhibitor is inhibiting a growth factor on the tumour cells or a survival factor made by macrophages?

Ristimäki: Angiogenesis seems to have popped up in almost every talk here. I recall that there is a paper suggesting that perhaps circulating endothelial cells could contribute to angiogenesis (Asahara et al 1999). Has anyone considered whether chemokines can attract circulating endothelial cells to sites of tumours?

Pepper: I don't think there are any data on chemokines in this process, but the idea that there are circulating precursors is being borne out repeatedly. People are finding that the precursors are there, and that they can be mobilized from the marrow by VEGF and GM-CSF. The problem is knowing to what extent they contribute to the formation of new vessels. For example, if you did a bone marrow transplant and you had labelled cells in the bone marrow, you would inevitably find some labelled cells in newly formed vessels. The issue is what percentage of new endothelium is derived from the initial vessel and what percentage is derived from the circulating precursors. Even more important is whether those labelled cells in the new vessels are endothelial cells which have incorporated into the vessels, or are they other cells which have come from the bone marrow which stick to the endothelium? It is difficult to evaluate to what extent these circulating precursors have an effect.

Gordon: Remember that there is a whole group of macrophages which was originally part of the reticular endothelial system. These are macrophages: they are not true endothelial cells, but they would fit with what you are talking about. In the liver there is a huge number of these.

Pepper: Another issue is that many angiogenic cytokines have been shown to up-regulate adhesion molecules on endothelial cells. Although we are thinking only in terms of angiogenesis, we should also think in terms of increased inflammatory cell recruitment in response to these angiogenic cytokines. By inhibiting angiogenesis you are not only inhibiting oxygen supply to the tumours, but you might also be reducing inflammation in a broad sense in the tumours.

D'Incalci: How can we do this? I think the only way would be to take several biopsies. This is impossible in many cases, but there may be some instances, for example when treatment is performed before surgery; that is becoming more and more common in the management of several tumours. More research should be done to discover and develop imaging techniques suitable for tumour biology characterization: this is the future.

Balkwill: It is easier if you are looking at macrophages.

Gordon: The resolution of these non-invasive methods isn't adequate yet.

Pollard: I'd like to comment on the issue of model selection. This is obviously a vexed issue: what is the appropriate animal model for a human cancer. As we know, traditionally transplantation models have been popular. My own feeling is that this is an approach that should pass. In many situations there are problems of clonal variability, and looking at a host versus graft response rather than an antitumour response. With the advent of sophisticated tools for mouse genetics, we really should try to move to more naturally occurring tumour types, driven perhaps by oncogenes that are relevant to the human cancers. The models I know best are breast cancer models, which have largely been generated by Bill Muller at McGill in Montreal. Bill's approach was to take powerful oncogenes such as polyoma middle T and also oncogenes such as HER-2/neu, and use these because they seem to be relevant to the human cancers. With HER-2/neu in particular, which is a signalling molecule, he then mutated various sites to see what signalling pathways are required for various aspects of tumorigenesis. These models are also very useful for preclinical trials. The question is, how relevant are they to the human disease? Mouse is different from human. We have looked at the polyoma middle T model closely: it does recapitulate many aspects of human breast cancer. We have done slide swapping tests with human pathologists and they can't tell which is mouse or human. When we look at this model, however, the stromal structure is quite different between a human breast cancer and a mouse. But these are much better models than transplantation of tumours. Now people have generated increasingly good models for melanoma, and are getting better models for prostate cancer. I think there will be a whole panoply of mouse genetic manipulations which will give you different cancers. The other good thing about them is their reproducibility. You can also use these to map modifier genes and pull out other genes that would influence the rate of tumorigenesis. This might then be translatable back into human. Then you have the problem of cost. The reason we chose the polyoma middle T model was because of reproducibility and speed. If you have a model like HER-2/neu which doesn't come down with cancer until 26–52 weeks, each experiment becomes a huge endeavour. The other advantage of the genetic models is that you can introduce other genes in a regulated way, and manipulate them. I'd like to push the idea of using the binary systems, for example, to be able to turn genes on and off. You can ask specific questions about functions of genes at different times. If you think that VEGF is an important molecule for tumour progression, there is no reason why you can't remove that from the tumour at different stages by using Cre–Lox systems. By using a mix of regulated systems and floxing of genes you can go a long way manipulating molecules *in vivo*. This is not being done as one might have imagined it would have been. If one thinks IL10 is important, for example, there is no reason why you can't remove IL10 or the IL10 receptor at specific stages of tumorigenesis using this technology. The traditional carcinogen-induced skin cancer models are probably quite good to study in this way because they more reflect what is going on. So I think we should all be doing mouse genetics, and then translate what we have back to humans. I should add that it is possible to humanize bits of the mouse genome.

Balkwill: I completely agree, but I would be a little more pragmatic about this. It is a question of time and money. If you just choose to use one model, you could argue that this is just one patient. I don't think we should totally throw the transplantable models out. In our lab we use c-neu transgenic mice as a genetic model of breast cancer, we use skin carcinogenesis models and we use ovarian cancer xenografts. I have shown the pathologists tissue sections of our xenografts and they think they are from a human patient. Also, some of the transplantable models show good features: the 410.4 breast cancer line has a leukocyte infiltrate that is not dissimilar to that in the human disease, and is there in about the same proportion. These genetic models are good but the experiments take a long time. Sometimes it is quicker to do a clinical trial than a mouse trial, and not much more expensive! We should use all the different types of models.

Pollard: I wasn't saying that we should use just one animal model. With the breast cancer models we need to use several different variants of them. I agree that you have to use some transplant models. There are only so many manipulations you can do in a mouse. Once you have three or four alleles the breeding becomes horrendous, and most of us don't have institutes in China to run these mouse houses. If you want to manipulate genes rapidly you need to use transplantable models. There have been quite nice models set up for both breast and lung cancer. In lung cancer, Roman Pererz-Soler takes the primary tumour and puts them straight into nude mice. They have 65% take. They can freeze the tumour and they can also freeze cell lines from the tumour and then they can transplant these to the lung. Then they do a drug test and use this to predict what is happening in the human. It seems to be quite predictive. Elizabeth Anderson from Manchester is doing the same thing with breast cancer, putting breast cancer biopsies straight into nude mice without any manipulation in culture. She then examines the effects of hormones, and these results have been impressive.

Balkwill: When you are doing this transplantation you are taking stromal cells and the whole structure with it. In ovarian cancer there is a nice model in which genetic lesions are introduced into primary normal ovarian epithelium. This results in a transplantable model with the appropriate genetic defects.

Pollard: Genetics gives us truth. It is powerful. If you can manipulate a gene in a cell type you can get a powerful result. I'm reminded of the talk by Louis Parada, who identified NF1. This is a Schwann cell disease, but when he made a knockout of the gene in Schwann cells so that they were null but the stroma was wild-type, he didn't get any tumours. When he removed one allele of NF1 from the stroma then the tumours came back. In this case the stroma and mast cells were the regulator of whether there was a tumour or not. This is an extremely powerful result that gives insight into a whole set of problems.

Mantovani: Turning to the chemokines, I would make four summary points from this meeting: recruitment, more than recruitment, metastasis, drug targets. There is a consensus that chemokines are the main driving force for the recruitment of leukocytes to tumours, although one should not forget other molecules such as VEGF or M-CSF. What is still unclear is the fine tuning of recruitment. Why don't we get neutrophils in tumours? In the genetic models there is good evidence showing the importance of early PMN recruitment. However, in the established clinical tumours there are generally no neutrophils, but there is plenty of IL8 and lots of CXC chemokines. We would expect these to recruit neutrophils in addition to angiogenesis, but we don't see this. We still don't understand the fine details of the microlocalization. In the tumours that we have looked at the dendritic cells and macrophages are differentially distributed, and we don't understand what is guiding this. The second point is that chemokines do more than just induce migration. They activate the genetic program, which includes TNF under certain conditions. We need to put this into a context. We also need to know what the genetic program of the leukocytes is *in situ*, and how this compares with that activated by IL10. The third point relates to metastasis. It is logical that tumour cells use the same tools that the professional migrating cells use for targeting their migration. In a way, it was expected to find chemokines involved in the process of metastasis. The worry is that some of the signals, such as CXCR4 ligands, are so widely expressed, that we don't have a molecular basis for the selective localization of tumours in different anatomical sites. There is a question as to whether these will ever become targets for therapy. Chemokines are attractants, but there is an intriguing story about what has been termed as 'fugitaxis': namely that high concentrations of chemokines can actually push out cells. In the thymus high concentrations of SDF1 promote the exit of thymocytes from the thymus. The properly controlled chemotaxis versus chemokine experiments with the proper analysis have not been done, but if this is true it would be relevant in conditions such as ovarian cancer. Paradoxically, this may be a way for the tumour cells to get out of the tumours. The final point concerns drug targets. These may be different in different tumours. My view is that collectively the data suggest that CXCR2 could be a valuable drug target in melanoma. There are clinical data, there is evidence from genetic manipulation and there is evidence in humans. Perhaps we have a drug target here. I also have a more general point. I am not as optimistic as Jacques Banchereau, but I am probably more optimistic than Siamon Gordon. I am not sure whether we are at a turning point, but much of what we are saying about inflammation is being tested in the clinic. Anti-TNF and anti-COX-2 are now in the clinic and soon chemokine inhibitors will be also.

Strieter: With regards to why neutrophils aren't in tumours, there have been genetic approaches to look at this. When IL8 is overexpressed in mice, there clearly is an attenuation of mouse neutrophils to traffic. This turns out to be related not to desensitization of the receptor but is mainly related to L-selectin shedding. This was published in 1994. Our paper published in 1996 taking xenografts of tumours in SCID mice clearly showed serum levels of IL8 that would rival the levels found in the transgenic mice. This is a unique situation in which the tumour has the opportunity to usurp neutrophil recruitment. In contrast, we have also published on adrenal cell carcinoma in which the hallmark of this cancer was significantly elevated levels of ELR-positive CXC chemokines. This tumour was highly associated with neutrophil infiltration. Not all clinical tumours are associated with the absence of neutrophils, but certainly the ones we have worked with have shown a paucity of neutrophils.

References

Asahara T, Masuda H, Takahashi T et al 1999 Bone marrow origin of endothelial progenitor
   cells responsible for postnatal vasculogenesis in physiological and pathological
   neovascularization. Circ Res 85:221–228
Leoni F, Zaliani A, Bertolini G et al 2002 The antitumor histone deacetylase inhibitor
   suberoylanilide hydroxamic acid exhibits antiinflammatory properties via suppression of
   cytokines. Proc Natl Acad Sci USA 99:2995–3000

Concluding remarks

Siamon Gordon

Sir William Dunn School of Pathology, University of Oxford, South Parks Road, Oxford OX1 3RE, UK

I think it was an ambitious goal to try to deal with two complex subjects and find a simple answer. But we have had some very useful discussion. I'd like to summarize a few of the themes that have emerged over the last three days.

First, are infectious diseases good models for tumour–host interactions? It is interesting to start thinking through the differences and the similarities. For example, in both cases there are two genomes with selective pressure on each but with different growth rates. If you think of it from the point of view of the organisms, we should aim at symbiosis and parasitism rather than host death. In the tumour there is a good coexistence for a while and unfortunately this may end in tragedy. From the perspective of a pathogen rather than a commensal organism, there are problems of invasion and evasion, and it is a highly dynamic system. Each is able to manipulate the other. So some of the paradigms that come from studying infectious diseases are quite useful.

There are clearly some key differences. One is the genetic instability in the case of the tumour and the relative genetic stability in the pathogen. Then of course the tumour doesn't want to spread in a population of individuals beyond the host, so it lacks the additional strategies for spreading outside the body found in infectious agents. Then there are constraints of time as well as space. The microenvironment is critical: every organ is different, not only in terms of oxygen tension but also in the many organ-specific factors we haven't yet come to grips with. The host does try to limit some of these invasions, for the tumour as well as the pathogen. It may try to isolate or encapsulate tumour cells. The misnomer of 'granuloma' which is not a tumour shows that there are some similarities.

The second theme of inflammation as a host response clearly is very varied, from the hot type inflammation we don't really see in these chronic diseases to much colder forms that are more like atherosclerosis and tumour biology. This depends on the cells and mediators involved. The idea that wound repair is a prototypical model has some value, but we must remember that you can repair most wounds and fractures perfectly, and it is very rare to find neoplasia as a consequence of this. This suggests that the regulation is tight. It will depend on factors such as chronicity and non-degradability or destructibility of a particular insult. We shouldn't get carried away by the idea that these are frequent complications of natural processes that are very well controlled.

One of the problems of inflammation is that it is not just one cell type we are dealing with. Many white cells, vessels and somatic tissues come into the act. I still think that one of the key problems is that of recognition. We have cells that have incredible effector capacity, such as neutrophils, which are pre-loaded. Yet they don't know that a possible target is there, or they are not triggered appropriately. We know nothing about recognition by the host cells and some of the pattern recognition molecules that may or may not be involved. These are receptors for modified self structures, and sometimes a molecule like the type A scavenger receptor can see apoptotic cells and it can see a bacterium. We don't really understand how it knows the difference between these two. In the case of apoptotic cells the host clears them without making inflammation by suppressing inflammation. PGE2 and IL10 are involved here. I'm beginning to think that macrophages are not pro-immunogenic cells: they are actually designed to prevent autoimmunity, and they have powerful suppressor activity in this regard. It is a bit like the dendritic cell which can also be suppressive, but it has a specialized ability to overcome this and to produce an immune response. The same is true of T cells. All of these cells can work both ways. I think the primary function of macrophages is to prevent immunity by efficient clearance of potential antigens. Finally there are other cells and molecules which we really haven't done justice to in terms of this recognition problem. The alternative complement pathway is critical, and then there are NK cells and $\gamma\delta$ T cells — all important systems which we haven't had time to discuss.

What are the possible roles of the inflammatory cells? In initiation, we haven't heard much to suggest that their genotoxic activities might contribute to DNA changes. Most people forget that macrophages are like wandering hepatocytes. They have inducible oxygenases and haem-converting enzymes. These cells have the capacity to metabolise carcinogens, and this is a neglected field. I was pleased that we heard epidemiological evidence that infections such as schistosomiasis and viral hepatitis increase the risk of tumours afterwards. In terms of promotion and pathogenesis, we heard many examples. We probably haven't paid enough attention to the surface molecules. I think the adhesion molecules are not just accessory to chemokines: there are some independent ones that may in turn control the chemokines as well, and production of other cytokines. What is very important for a chronic process like this is persistence of the stimulus. When it comes to metastasis and invasion, we have heard several examples of the roles of proteases and chemokines, but I think we are forgetting about the intrinsic ability of cells to move. Migration of cells is a fundamental property. The matrix is another subject we haven't done justice to. It can bind growth factors and chemokines, and can influence many aspects of gene expression in white cells as well as tumour cells. Furthermore, there is the question of heterogeneity of the matrix. There is almost no literature about how the matrix in different tumours is different from that in the connective tissue or bone. This is a huge gap. When it comes to migration of leucocytes, the paradigms we have of systemic and local cell migration are very powerful. For example, the splenic marginal zone is highly specialized to capture circulating cells. This is an interesting physiological process which may be relevant to the way tumour cells might be captured from the circulation or lymphatics. What about the stroma? 'Stroma' means 'bed', and it is a bed in which you lie. The haematopoietic stroma is probably the best characterized, but there is still a lot of ignorance about how these particular fibroblasts are specialized. Macrophages are also part of the bone marrow stroma. When you come to tumour stromas we have the same type of complexity. We all recognize its importance, and it would be nice to try to put some of these different systems together. We didn't hear much about bone marrow: do cells which metastasise to bone marrow home just because of the ligands for chemokines? I find this hard to believe. I think there are adhesion molecules in each of these sites, and we know some of them are candidates to attract cells from a circulating pathway. I asked why the spleen does not usually get metastases, and heard of one exception, but there is a huge blood flow through this filtering system.

How do we study this? Jeff Pollard has given us a nice account of mouse models, but we are ignoring simpler animal systems. *Drosophila* has taught us about innate immunity and development, and *Caenorhabditis* has taught us about apoptosis. I don't know of anyone studying aspects of this process in such simple model systems. The crown gall tumour in plants is a fascinating example. Zebrafish may give us nice models for looking *in vivo*. Between mouse and human there are some species differences we shouldn't forget about. I like the idea of making an observation in human, taking it to a mouse and then coming back to human. The CSF story was part of this, and there may be other examples. The human tumours do offer us a lot of natural history and it would be a pity not to use this as well as we can. The human polymorphisms may give us more insight than we already have.

What about the cell biology and histology? I think it is very hard to recover these cells from the embedded environment. We have tried, and it is tough going. It is necessary to look at them *in situ* with good antibodies. We have seen some nice examples here. I also like the idea of looking at the proteins in tissue arrays with good reagents. We are looking for changes that are predictive of behaviour. I like Jeff Pollard's needle experiment of pulling out cells. We are told when we sample these tumours for array studies we have to be careful not to leave out the stroma. Where we sample is also important, because human tumours are much bigger than mouse tumours. I also like the idea that the tumour cells are not only expressing genes that are aberrant. The differentiation and growth are wrong, a lot of the signalling pathways are turned on and they produce molecules that they normally wouldn't.

We clearly need more surrogate markers so that we can see there is something wrong in the bloodstream of the tumour bearing host. With regard to host responses, we need the equivalent of the acute-phase response, something that is chronic and inducible. There must be something here that we haven't exploited. Then, when it comes to treatment, we could target drugs to malignant cells more selectively. We could even make pro drugs that are then processed by the host inflammatory cells. Then there are the monoclonal antibodies and the inhibitors we have heard about today. I think we need a lot more basic knowledge, and we shouldn't be too quick to look for the answers before we know what the questions are. I think this has been a great meeting and I would like to thank you all for your contributions.

Index of contributors

Non-participating co-authors are indicated by asterisks. Entries in bold indicate papers; other entries refer to discussion contributions.

Subject index